I0477480

Consensus Solutions

Perspectives on Monetizing the Deepwater Gas Reserves in Niger Delta, Nigeria

Consensus Solutions

Perspectives on Monetizing the Deepwater Gas Reserves in Niger Delta, Nigeria

by

Funmi R. Ebiwonjumi

Copyright © Funmi R. Ebiwonjumi, 2018

Table of Contents

List of Tables

PREFACE

This book is a response to an opportunity I had to lead a team of eleven subsurface geoscientists from a consortium of multinational oil companies to assess the quantity of hydrocarbon resources in a straddling acreage in deepwater Niger Delta to recommend ways to harness the resource. The result of the study indicated enormous gas deposits with comparatively smaller quantities of oil. The contract for extraction of hydrocarbon from deepwater Niger Delta did not include gas deposits. The path towards monetization of the deepwater gas is locked due to a lack of government policy despite Nigeria's need for the gas as a developing nation. Problems with the extraction and sharing of oil wealth from the Niger Delta over the years made the quest to seek consensus solutions to monetize the deepwater gas resource a worthwhile challenge. My 25 years of professional oil and gas subsurface expertise and working knowledge of six different world basins may contribute to Nigeria's development at a national level and foster more effective integration of different stakeholders' perspectives to arrive at consensus solutions for monetizing Niger Delta deepwater gas.

This study of monetizing deepwater gas deposits in Niger Delta, Nigeria resulted in better understandings of the perspectives of three key stakeholders: members of the community, government agencies, and oil industry experts. The research sample included 27 stakeholders from these groups who completed a survey and participated in focus group discussions. They had sufficient knowledge about deepwater gas deposits and the lack of government policy concerning deepwater gas monetization. Most participants reported that government policy on sharing oil wealth was not satisfactory; the government received the most financial benefits and communities received the least deepwater oil and gas proceeds. Gas extraction from deepwater may increase power and steel development, fertilizer and petrochemicals plants, and growth and industrialization in the Nigerian economy.

Study participants believed that environmental and health hazards due to gas flaring during onshore oil and gas extraction would be minimal during deepwater gas exploitation due to the absence of immediate community settlements next to proposed deepwater gas extraction facilities. They suggested a policy providing incentives for deepwater gas exploitation to enable more accurate determinations of gas reserves in deepwater Nigeria. Additional suggestions included provision of adequate security during deepwater gas extraction, which could curb criminality, vandalism, communal clashes, and corruption in the Niger Delta. Stakeholder theories informed the research findings. Existing multinational organizations' safety standards and sustainable community development efforts reflect corporate social responsibility (CSR). A systematic examination of thematic results from the different stakeholder focus groups presented a consensus on specific policy proposals and changes of law concerning deepwater gas deposits in Niger Delta.

Nigeria has been a steady producer and exporter of oil since the early 1960s. Before this time, Nigeria was a prosperous agrarian society with agricultural exports that provided the main foreign exchange earnings for the federal government. Infrastructural development fostered a sustainable agricultural economy, and road and rail networks supported the gathering and transportation of products for export. The Niger Delta province was rich in both oil and gas, but the exploitation of gas has been minimal. Oil was preferential even though abundant natural gas resources in Nigeria are larger than oil reserves. Hydrocarbon prospects of the exploration phase were risky if there was a preponderance of gas. Currently, Nigeria has about 187 trillion cubic feet of gas reserves distributed equally between gas associated with oil accumulations and non-associated gas reserves. This estimate will likely increase if governmental policy or economic necessity results in more gas exploitation in the region. Nigeria uses very little gas, mostly for domestic cooking. Industrial uses include the production of methanol, fertilizer, and electricity generation. The federal government of Nigeria lacks a gas policy that regulates the unrestricted development of the gas infrastructure with an appropriate legal, commercial, and regulatory framework. The federal government approved a policy on the implementation of renewables and energy efficiency in 2015 (Ministry of Power, 2016).

A substantial portion of the associated gas that comes out during oil production is flared. This is due to a lack of infrastructure for gas transportation through pipelines. Hazardous air pollutants emitted from gas flaring cause adverse health reactions in humans and damage the ecosystems through pollution. Gas flaring is also a major cause of environmental pollution and a waste of non-renewable resources that results in huge loss of revenue for the country. Gas crews began flaring during oil extraction in the 1950s in Nigeria and many groups pursued government policy to reduce gas flaring during projects. Companies face penalties for continuous flaring of gas during oil extraction operations. The government contributes to joint ventures (JVs) in oil extraction. In JV arrangements, the host government takes part in the concessionary system as working interest owners. A joint operating agreement

(JOA) is often drawn for the execution of the operations (Saidu & Sadiq, 2014).

The Nigerian liquefied natural gas (LNG) plants have up to six trains in production and more trains are under construction (Nigerian LNG, 2016). LNG companies boost the gas contribution to the national economy, but gas extraction only occurs onshore in the Niger Delta. LNG infrastructures are grossly inadequate to fully utilize the gas reserves in the Niger Delta onshore and completely unsuitable for deepwater. Nigeria has limited technology and an insufficient industrial base for energy consumption (Ukpohor, 2013). Gas deposits in deepwater acreage in Nigeria are under a different contractual term with international operators than onshore acreage, and there is no fiscal policy for effective exploitation. Companies only collect and channel associated gas that comes with oil production onshore for eventual monetization. According to the production sharing contract (PSC) agreement between the Nigeria National Petroleum Corporation (NNPC) and oil companies, the cost of deepwater exploitation is recoverable with crude oil in the case of a commercial find with provisions made for tax oil, cost oil, and profit oil.

The following chapters explore value drivers for monetizing deepwater gas reserves, which may contribute to national development in many ways. This book includes details of the history of the oil and gas industry in Nigeria and government policies to assure safe harness of these resources to benefit the populace. Sharing oil wealth between diverse entities with different political interests and economic powers is complex. The perceptions of key stakeholders regarding ways to secure their interests inform this quest to facilitate extraction of deepwater gas for utilitarian benefit. Strategies to arrive at consensus regarding how to monetize the deepwater gas reserves must reflect recommended actions from key stakeholders; these are the results of this action research.

CHAPTER 1

WHY MONETIZE DEEPWATER GAS RESERVES?

In the early 2000s, legal governance structures of the 1970s were no longer adequate for the Nigerian oil and gas industry. For example, though amended many times, the Petroleum Act of 1969 remains despite its design for the industry at its infancy. Similarly, the NNPC Act of 1977 is out-dated legislation that does not align with contemporary global business realities (NNPC, 2016). These laws span several pieces of legislation, numerous amendments, policy statements, and regulations in several documents that are often difficult to locate. The Ministry of Petroleum Resources remains a civil service organization that is ill-equipped to formulate policies for the complex, modern industry.

The regulatory body, the Department of Petroleum Resources (DPR), is similarly constrained as are national oil companies (NOCs). The Ministry of Petroleum Resources is a typical state institution that operates as a cost center with little or no sensitivity for the bottom-line. The need for far-reaching reforms of the country's oil and gas industry is evident.

The desire to address these issues gave rise to the National Oil and Gas Policy. The policy covers all relevant aspects of the industry: upstream and downstream, gas and petrochemicals. The new policy ensures separation and clarity of roles of different public agencies operating in the industry. The policy includes strict commercial orientation into all aspects of the industry. The Nigerian Petroleum Industry Bill (PIB) reformed the oil industry and revealed the guiding forces of local politics (Pérouse de Montclos, 2014). The PIB exposed the limitations of the state's ambitions, desire, and capacity for reform. The PIB strongly reflects the regional divisions and social tensions. The PIB's overall objective was to reposition the oil and gas industry in view of contemporary challenges to secure maximum sustainable value for the nation. This is especially important due to volatility of oil prices that often affects Nigeria's economy.

The Quest for Oil Industry Reforms

The National Oil and Gas Policy provided broad guidelines for the emergence of the new oil and gas industry, which became a concrete legal and institutional framework to transform the industry. According to Okoye (2012), delays and uncertainty in implementation harm the industry and the country. The federal government constituted a second Oil and Gas Sector Reforms Implementation Committee to implement the Oil and Gas Policy report. The committee advised new bodies, institutions, organizations, and agencies that constitute the institutional framework for the restructured oil industry. Focused leadership and reform of the public sector are critical to ensure that economic reforms are effective (Nwokoma, 2015). According to Dias-Sardinha and Reijnders (2001), the most common environmental objective of organizations is compliance with the law. Legislations and regulations must have sufficient attractiveness for the operators in the oil industry to choose to implement them. Appropriate legislation may sustain businesses in the future.

Policies that confront issues due to restiveness in the Niger Delta received little attention (Okpanachi, 2011). There are separate incentives for oil and gas extraction in the Nigerian state because the gas industry is much less developed than the oil industry. According to Kizito (2014), the Nigerian tax system requires reform to have a significant impact on economic growth. The government should create policies and programmes that enhance the income of citizens to accelerate consumption, investment, employment, and tax revenue.

Continuing the current mode of operation for the extraction of oil and gas will not result in the effective utilization of Nigeria's gas resources. The federal government has a responsibility to ensure equitable and efficient utilization of gas resources. The state and local governments must ensure equitable distribution of resources to their constituencies for sustainability. Gas is necessary to generate income for growth and development, and policies must support strategies to unlock the potential of gas to benefit the people of Nigeria. The mineral resources in Nigeria belong to the government, and the federal and state governments must implement programs for the utmost benefit of the public. The ability to reach governmental and public consensus on a strategy for profitable extraction by oil and gas companies requires understanding the perspectives of diverse groups of stakeholders.

Sustainable Development

According to Ibaba (2008), sustainable development strategies in the Niger Delta area have gaps that constrain efficiency. Attempts by multinational oil companies to implement community development projects have not fully captured the aspirations of the populace. Government attempts to develop the Niger Delta established various organizations to facilitate sustainable development. These bodies experienced problems with resource allocation and accountability and had no real impact on the public. According to Orogun (2010), economic exploitation of the region's vast crude oil reserves by multinational companies and government authorities resulted in environmental

15

devastation, excruciating poverty, and the recurrent rule of impunity. PSC arrangements govern the relationship between the Nigerian government and oil companies in inland deep and ultra-deepwater acreages. The agreement is that the contactors (i.e., multinational oil companies) bear all the risk. After discovery and development of oil fields, the companies recover their investment with profit from produced oil. The agreement is exclusively based on oil extraction and does not address gas reserves.

Deepwater gas is costlier to extract for use and export, and the government omitted it from PSC terms. The community of the Niger Delta felt frustrated after the failure of the Nigerian government to satisfy their socio-economic needs, which led to escalated violence in the region (Aminu, 2013). Deepwater activities in the Niger Delta will affect the local communities, much like the Macondo oil well blowout on the 20th of April 2010 off the coast of Louisiana that affected communities at the shore (Macondo, 2014). The need to utilize gas as part of the sustainable development of the nation is critical. Exploitation for domestic consumption, power generation, or export is most expedient with public participation. The present study's focus is on projects, policies, and programs that will most effectively benefit Nigeria's economy and people in a sustainable manner.

According to Hsu and Chang (2013), traditional cost-benefit analysis is insufficient in the energy sector. The energy sector requires evaluation of emerging energy technology, commercialization of target products, and valuable products generated in the process of producing target products. There is an essential need for deepwater gas reserves as viable assets that benefit all key stakeholders. This study built on the local knowledge of different groups of stakeholders to suggests ways to achieve much-needed change.

Need for Consensus

There are three main classifications of stakeholders: government, public, and business. Developing a consensus based on common ground thinking between the government, community, and companies may result in new approaches to meet the needs of all members of these three groups. The Nigerian oil and gas industry is a challenge for researchers. There is insurgency, media and international activism, bunkering, piracy, and kidnapping. There are also ecological challenges (e.g., oil spills and gas flaring, livelihood eradication, and institutionalized corruption). These issues affect public policy for commercial exploitation of deepwater gas resources in the Niger Delta. There are currently no explicit terms for deepwater gas monetization. Corruption in all facets of the Nigerian state affects the economy and livelihood of the people.

Despite more than four decades of oil export for revenue, there are few investments in the nation that share oil wealth with the Nigerian state. Instead, scandalous stashing away of stolen wealth is commonplace among those in control; corruption is the curse of development in the Niger Delta region. Therefore, a policy for agreement on monetization of deepwater gas will be a challenge. The researcher collected data from all relevant stakeholders to develop a consensus for mutual benefit. This study explored stakeholder perceptions to inform a commonly accepted strategy for the government, communities, and corporations. A consensus between sectors may define a new approach to policy on deepwater gas pricing based on the perspectives of government officials, community members, and corporate personnel. There is a need to utilize gas reserves in deepwater; Nigeria is currently underutilizing its gas potential. There may be significant job growth if deepwater gas companies invest in cutting-edge global extractive technologies. If policy can support an equitable approach to gas extraction, economic benefits may accrue in Nigeria, which would benefit states, communities, and oil and gas operating contractors. This research is significant because the environmental footprint of deepwater gas is much less than onshore gas; finding ways to move forward with deepwater gas operations may create environmental advantages.

The overarching goal was to discover the perspectives and opinions of members of three focus groups (the government, oil companies, and community) on deepwater gas reserves monetization. What is the level of knowledge concerning oil and gas resources and income generation? What is the level of knowledge in the government about policies to exploit and monetize deepwater gas for profit? What is the possibility of utilizing deepwater gas in the same way as onshore gas for domestic purposes? Does the government have an interest in giving incentives through a monetization policy to help harness deepwater gas profitably? Does monetization of deepwater gas help industrialization by empowering steel, fertilizer, and petrochemical industries? What is the effectiveness of the collective opinions of the three focus groups in influencing government decisions on a deepwater gas monetization policy? What are the benefits for the government of the global experience of multinational oil companies (such as Shell) in shaping deepwater gas pricing? Is there any benefit for the communities harnessing deepwater gas deposits? The book will explore each of these questions according to responses from key stakeholders.

The Search for Consensus

This study of monetization of deepwater gas reserves occurred over 20 months in 2016 and 2017. With a population of about 198 million people and inadequate infrastructures, the need for development in almost all sectors of the Nigerian economy is clear. The communities associated with the gas deposit are interested in protecting their environment from any form of operational degradation and livelihood eradication. The communities also wish for gainful employment for their community members. The researcher assumed that the contractor oil company was interested in responsibly extracting gas for profit without harm to people or the environment and that it is possible to develop a consensus-based strategy for generating new income from deepwater gas assets.

Limitations to this study included the logistics of accessing rural communities through swampy creeks to hold focus group discussions. This was difficult, especially due to security risks (e.g., kidnapping and insurgency). Another limitation was that changes in government could redefine operating conditions and terms to govern proposals for gas commercialization policies in the deepwater, which could result in a lack of interest in monetizing the deepwater gas. According to Oduntan (2008), it is difficult to ascertain the extent to which emergent governments can create environments for sustainable exploitation of the resource-rich deepwater Niger Delta. The passing of the PIB modernized procedures in the oil and gas industry in general and increased interest in deepwater gas pricing regulation, which made consensus regarding monetizing deepwater gas difficult.

There was also a language barrier when conducting research of this nature with community members who only spoke local dialects. Many activist groups oppose oil and gas exploitation in the Niger Delta because they feel current policies do not adequately represent their interests. The rise of militarism and terrorism in the Niger Delta was the result of the federal government and oil companies' obstruction of non-violent protests for environmental justice in the Niger Delta (Madubuko, 2014).

The sustainable development of oil and gas resources for the benefit of key stakeholders is difficult and there is a dire need for consensus regarding monetization of natural resources. The following chapters include the history of oil and gas exploration and exploitation in Nigeria and government operating agreements to ensure safe and environmentally friendly extraction and equitable distribution of oil and gas wealth. Many stakeholders are dissatisfied, as became clear during focus group discussions. A consensus approach is a noble solution that may come at its own

heavy price.

CHAPTER 2

HISTORY OF THE NIGER DELTA OIL AND GAS INDUSTRY

Nigeria is the world's eighth most prolific exporter of crude oil; the industry accounts for over 40% of Nigeria's gross domestic product (GDP) and over one billion dollars of investments annually (Atsegbua, 2012). Much of Nigeria's history is the result of oil extraction and its effects on communities, politics, government agencies, financial institutions, media, and various other ancillary organizations. Little past research focused on monetization of the deepwater gas reserves in the Niger Delta. Yet, gas is increasingly important in the world as a source of cleaner fuel for both vehicular and industrial consumption. Scholarship concerning Nigeria's public policies for monetization of deepwater gas reserves in the Niger Delta is minimal because drilling companies hold virtually all the data upon which policy is based. Discovering stakeholders' perceptions of monetization policy requires understanding the social, political, and economic environments within which the policies exist. It is therefore necessary to consider the socio-economic environment and political conditions in Nigeria.

Nigeria and the Niger Delta Region

Nigeria is a Federal Republic in West Africa; its southern coast is on the Gulf of Guinea in the Atlantic Ocean. Nigeria includes 36 states and a federal capital territory, Abuja, which is the seat of government. Before the arrival of British colonialists, tribes governed the territories of what became Nigeria. Modern-day Nigeria originated from the merging of northern and southern protectorates in 1914 by the British. These administrative and legal protectorate structures created indirect rule through traditional indigenous peoples' institutions. Nigeria gained independence from the British in 1960 and became the Federal Republic of Nigeria in 1963. Nigeria remains the most populous country in Africa, the seventh globally, with an estimated population of over 198 million (National Population Commission [NPC], 2018).

The geologic separation of the South American and African plates over the last 252 million years at the triple junction created the space that is the modern-day Niger Delta. It includes a mix of source and reservoir rocks as the Benue and Niger Rivers debouch into the Atlantic. The states of the Niger Delta region include: Abia, Akwa Ibom, Bayelsa, Cross River, Delta, Edo, Imo, Ondo, and Rivers. Their population is only about a quarter of the Nigerian population, but the states have a diverse set of ethnic groups and dialects spreading through about 5,000 communities. Oil is the mainstay of the Nigerian economy. According to Obi (2009), the Niger Delta states are the oil producing states of the federation.

The people of the Niger Delta region have a rich and diverse cultural heritage with over 40 different languages and dialects. The region is inhabited by minority ethnic groups of Ijaw, Urhobo, Efik, Ibibio, Ogoni, Itsekiri, Edo, Yoruba and Igbo, with a population of 32 million people (Madubuko, 2014). According to Edino, Nsofor, and Bombom (2010), several questions need asking regarding perceptions and attitudes towards environmental problems in the Niger Delta area where political tension and economic adversity are prevalent. The intermingling of political and economic interest prevents a holistic search for answers in the environmental and socio-economic crisis that has engulfed the region.

From the very beginning of oil exploration in Nigeria in 1937 until early 1993, virtually all exploration and production occurred on land, swamps, and shallow offshore waters. Initial offshore exploration was an extension of the onshore successes in a shallow water region. The federal government of Nigeria opened a new frontier in oil and gas exploration in 1993, heralding a promising economic future. The PSC arrangement governs the understanding between the NNPC and all new participants in the new deepwater acreages (NAPIMS, 2016). The government allocated some offshore blocks in water depths reaching 2,500 meters (i.e., deepwater). These were technically challenging and capital-intensive areas for oil and gas operations because no infrastructure was in place. Nonetheless, during the first decade of drilling in deepwater, companies made several major discoveries in stacked deepwater turbidite reservoirs. The reservoirs are of high quality; fluid presence in the reservoirs show up acoustically in seismic data. Niger Delta deepwater reservoirs required cutting edge geophysical prospecting techniques. These techniques revealed a string of medium sized oil, gas, and condensate fields along the hydrocarbon rich structural and stratigraphic traps of the Miocene age.

New economic potential may result from growth in the resources in the deepwater Niger Delta. Work beyond the inboard side of the outer fold and thrust belt involves more exploration risks and greater challenges for prospecting of oil and gas. Potential risks include: source rock and its maturity, migration of hydrocarbons due to fault detachment that act as barriers, and the timing of trap formation. Crews use imaging tools to explore hydrocarbons at deeper depths in areas already in production. New, deeper depths require preparation to work in high pressure environments. There is also a risk of encountering mostly gas at deeper depths and the production sharing agreement does not expressly state extraction terms for gas.

History of Oil in Nigeria

Almost the entire country of Nigeria was granted to Shell D'Arcy in the 1930s to explore for oil. The company struck oil in commercial quantities in 1956 at Oloibiri, a small remote creek community in present day Bayelsa State. Following this discovery, Shell increased exploration activities and discovered oil in eleven more areas in the Niger Delta by 1958. The Oloibiri field began production in late 1957 with ten wells (six appraisal wells and four development wells). Shell played a dominant role in Nigeria's oil and gas operations until 1971 when Nigeria became a member of Organization of Petroleum Exporting Countries (OPEC). Firmer controls of the country's oil and gas resources began with the emergence of NOCs across OPEC member countries. The main objective of the NOC was to monitor the stake of oil-producing countries in the exploitation of their nation's resource. Oil was a major source of income in most oil producing countries; monitoring the income and expenditure from oil was a critical part of national accountability. In Nigeria, the prudent utilization of oil revenues remained elusive for policy-makers over time. Degeneration was extensive and successive governments struggled to effectively account for their predecessor's oil revenue deployment. This created great disaffection between the federal government and the expectation of other beneficiaries who have equal stakes in the utilization of oil wealth.

Nigeria has huge reserves of oil and gas, which are mainstays of the economy. The quest to produce these resources made the landscape competitive for multinational and indigenous oil and gas companies. An income inequality followed natural resource booms in Nigeria; inequality falls immediately after a boom and increases steadily over time until the initial impact of the boom disappears (Goderis & Malone, 2011). Therefore, natural resource theory applies in the Nigerian state; the nation seeks to maximize value from deposits of natural resources within territorial waters. Public choice theory stipulates that a rational, self-interested individual will seek maximum utilities by employing any means to achieve the greatest benefit at the least cost. The present research explored the sensitivity of individuals to public choice theory for personal benefit.

Nigerian Oil and Gas Regulatory Agencies

Many governmental entities regulate the operations of the oil and gas industry in Nigeria. Prominent amongst them is the Ministry of Petroleum

Resources, which represents the administrative arm of government that formulates policies and provides general direction to other agencies in the sector for exploration and production of the nation's oil and gas resources. According to Ekhator (2015), the state-oriented regulatory regime of the Nigerian oil industry is ineffective. The ministry also provides oversight to other sectors in the industry: midstream, downstream, and services with a federal minister. Inadequate oversight could lead to shortcomings rooted in policy imperfections, a weak regulatory regime, organizational deviance in lieu of integrity, and inter-organizational structural deficiencies (Kurtz, 2013). Some of these elements plagued the Nigerian oil industry with far-reaching effects on the people and environment.

The NNPC is a statutory corporation through which the federal government participates in the oil and gas industry. The NNPC's primary function is to oversee the regulation of the oil industry with secondary responsibilities for upstream and downstream development. The NAPIMS is the subsidiary of NNPC that supervises the Nigerian government's investments in the oil industry. Investments from the upstream segment to midstream and downstream gas utilization are key areas of coverage for NAPIMS. NAPIMS also maintains assurances of CSR and effectiveness of health, safety, and environment activities of the companies. According to Abiola and Ashamu (2012), NNPC managers are aware of environmental accounting practices and actively deploy financial and environmental strategic business units.

The DPR is responsible for ensuring compliance with terms governing the award of oil licences to companies engaged in petroleum operations. Other functions include monitoring oil companies' operations to ensure consistency with international industry standards and practices. DPR also regulates issuance of annual permits to oil and gas companies, without which they would be unable to operate in the industry.

The Nigerian Investment Promotion Commission (NIPC) is responsible for registering foreign investments in Nigeria (NIPC, 2016). NIPC also acts as a liaison between investors and government ministries, departments, institutional lenders, and investment groups. The NIPC grants incentives to woo investors (e.g., enlargement of modes of payment for foreign equity to include spare parts, raw materials, and other business assets without initial disbursement of foreign exchange from Nigeria). This guarantees smooth transfer of dividends and profits from foreign investment in Nigeria and capital repatriation in the event of liquidation. Dividend payments are subject to withholding tax at 10% as final tax. Aigboduwa and Osaimoje (2012) emphasized access to funding as a means of encouraging participation of small and medium scale entrepreneurs in the capital-intensive oil and gas industry.

The National Maritime Administration and Safety Agency (NIMASA) monitors and promotes the development of indigenous and commercial shipping in international and coastal shipping trade. They regulate and promote maritime safety, security, and marine labour (NIMASA, 2016). The Nigerian Content Development and Monitoring Board (NCDMB) implements the provisions of the Nigerian Oil and Gas Industry Content Development Act of 2010 and coordinates, monitors, supervises, administers, and manages the development of Nigerian content in the oil and gas industry (NCDMB, 2016). Nigerian content policies have the potential to succeed where previous policies failed to translate resource wealth into economic and social development (Ovadia, 2013).

NCDMB (2016) also assists local contractors and Nigerian companies to develop capabilities and capacities. The key areas of focus include: training and employment of Nigerians; facilitating the establishment of critical facilities such as pipe mills, docking, marine facilities, and pipe coating facilities; promoting indigenous ownership of marine vessels and offshore drilling rigs; integration of indigenes and businesses residing in oil producing areas into mainstream industry economic activity; and promoting services that support industry activities such as banking, insurance, and legal work.

The Niger Delta Development Commission (NDDC) formulates policies and guidelines for the development of the Niger Delta area. The commission conceives, plans, and implements projects in accordance with set rules and regulations for the sustainable development of Niger Delta transportation; prepares and estimates costs of plans to promote the physical development of the Niger Delta area; and implements all measures for the development of the Niger Delta area by the Nigerian government and the member states of the commission (NDDC, 2016). The NDDC identifies factors that inhibit the development of the Niger Delta area and assists in the formulation and implementation of policies to ensure sound and efficient management of the resources of the Niger Delta area. The NDDC is also responsible for ecological and environmental problems. The integrity pact in the NDDC reinforces and complements existing anti-corruption initiatives in Nigeria's public sector (Idemudia, Cragg, & Best, 2010). The commission advises the federal government and member states on the prevention and control of oil spillages, gas flaring, and environmental pollution. The federal government of Nigeria approved the National Oil and Gas Policy on the 5th of September 2007. The singular objective of the policy was to make far-reaching changes to ensure the fundamental transformation of Nigeria's oil and gas industry to meet requirements of 21st century global industry standards.

CHAPTER 3

GOVERNMENT POLICIES AND
PRACTICES IN THE OIL
AND GAS INDUSTRY

In developed democracies, the rule of law is paramount and public policy processes distribute authority and resources among government and quasi-government organizations. Effective public policy requires that government agencies, partners, and collaborators influence proprietary and non-profit sectors. These distributions reflect a wide variety of political and technical considerations that vary by context. Lynn and Robichau (2013) found that the core influences on government performance are hierarchically ordered structures (i.e., organizations, delegations of formal authority, rules and guidelines, categorized budgets, information exchange, and reporting requirements) and operational mechanisms that enable and constrain public administrators in their policy implementation roles. The Nigerian government shares oil wealth such that every part of the economy benefits from the oil resource. Great dissatisfaction exists between the different sectors, which led to political upheavals and disruption of fragile peaceful coexistence in the culturally diverse nation. Policy makers protect their own interests before adjudicating the rights and privileges of their constituents.

Hilmer (2010) stated that democracy involves open and public, relatively egalitarian, informal political practices beyond institutional boundaries, rational deliberative processes, and instrumental actions of representative government. The renewal of participatory democratic theory as a distinct theory of democracy may spur political scientists to focus on undervalued aspects of democratic participation. Participatory democracy is not necessarily more progressive, inherently more valuable, or generally superior to other forms or theories of democracy. A central tenet of participatory democratic theory is that citizens who actively participate in self-governance experience a heightened sense of political efficacy and empowerment. Such democracy should express the values and interests of the politically marginalized, as seen in the Niger Delta area. In participatory democracy, citizens become educated, effective, and empowered because they continually exert a high level of control over the social, political, and economic institutions that directly affect their lives (Hilmer, 2010). The oil-rich communities of the Niger Delta have little say in the distribution of oil wealth by successive governments.

Operating Agreements

Nigeria did not set up competitive NOCs like some of the OPEC nations. Instead, continued operations of the international oil companies (IOCs) with JV agreements specify responsibilities of the Nigerian government and the IOCs. Shell was initially the predominant IOC. Other companies also operated under JOAs with the NNPC with different stakes in different acreages. According to Atsegbua (2012), there was a marked absence of indigenous players involved in oil and gas transactions; the industry imported over 90% of goods and services from overseas. The passage of the Nigerian local content bill was a significant development for domesticating the oil and gas industry through local value additions to the local economy. The law established the Nigerian Content Monitoring Board that manages the coordination, monitoring, and implementation of local content law.

Previous acts of the government tried to develop a local content framework for the industry without success. The local content law increased indigenous participation in the industry by prescribing minimum thresholds for the use of local services to promote employment of Nigerians. Compared to Saudi Arabia, Venezuela, and Kuwait, the local content law greatly empowered indigenous oil and gas companies to assist Nigeria in developing the technical capacity for the industry.

After three decades of exploring onshore, the Nigerian government conceded to exploration in the offshore blocks of the Niger Delta. The policies for onshore oil and gas exploration by the Nigerian government did not address the technically challenging deepwater environment. Therefore, a shift in governing guidelines was necessary for offshore oil and gas operations. The move from a JOA regime to PSCs changed the operation and regulation of the oil industry in Nigeria. This shift resulted from the complexity of operations in the offshore terrain that makes regulation under a JOA more difficult and from the dwindling resources of the country that decreased funding under the JOAs. The Nigerian government wanted to increase oil and gas reserves, but funds were not available; hence, the PSC offered a funding arrangement.

According to Ogbonna and Ebimobowei (2012), the abundance of petroleum and its associated income was beneficial to the Nigerian economy between the years 2000 and 2009; oil revenue had a positive and statistically significant relationship with GDP and per capita income, but its relationship with inflation was negative and not statistically significant. Similarly, petroleum profit tax (PPT) and royalties had positive and statistically significant relationships with GDP and per capita income, but the relationship with inflation was negative and not statistically significant. Licensing fees had a negative and a non-statistically significant relationship with GDP and per capita income respectively, but the relationship with inflation was positive and statistically significant. Income from the nation's natural resource had a positive influence on economic growth and development. Based on secondary data from Central Bank of Nigeria's statistical bulletin, Nigerian National Bureau of Statistics, and the NNPC, petroleum income (oil revenue, PPT, and royalties) positively and significantly impacted the Nigerian economy when measured by GDP and per capita income (Ogbonna & Ebimobowei, 2012). The successes of the PSC encouraged the Nigerian government to consider its extension to other areas of the industry that operated under JOAs. The shift from JOA to PSC as a contractual model in the Nigerian oil and gas industry encouraged investment in Nigeria.

Nigeria is dependent on oil and gas revenue as the main source of national wealth. According to Chindo, Naibbi, and Abdullahi (2014), oil featured prominently in Nigerian politics within various tiers of the federal government, particularly as it relates to principles for controlling and sharing oil wealth between oil producing and non-oil producing parts of the country. These decisions influence inter-ethnic relations and the distribution of power in the multi-ethnic federation. Escalating violence in the oil-rich ethnic minority of the Niger Delta region of Nigeria includes insurgency over the sharing of oil revenue. The recent crash in oil price required finding new ways to profitably exploit undeveloped gas reserves in the deepwater. Nigeria was extraordinarily dependent on the oil sector, which accounted for over 90% of exports and government revenues and one third of the GDP. Nigeria's resource wealth did not translate into meaningful development.

Nigeria's poor state of development may be a product of the pathologies known as the resource curse. The collective impact of corruption, government complacency, the Dutch disease, lack of public accountability, neglect of education, and excessive external debt/borrowing hampered the development of the country (Chindo et al., 2014).

Nigeria ventured into PSC arrangements after the Arab oil embargo of 1973. The venture contributed to a shift in the relationship between oil producing countries and their oil-consuming counterparts. PSCs became popular worldwide as producing countries sought to gain better control of the industry. They benefited the state more than traditional JV agreements and were good solutions for the federal government. Oluwatosin, Abimbola, and Olusegun (2011) noted the need to spend oil revenue productively for real output growth. Diversification efforts occasioned by distributive investment of oil revenue in other sectors served as a cushion for the government. Gains included financing the government's share of oil and gas JVs and stimulating a non-oil-dependent economy that could effectively weather oil price volatility. The enclave nature of Nigeria's oil sector with weak linkages to other sectors of the economy made it imperative to spend oil revenue productively to create output growth.

The main features of a generic PSC agreement include that the contractor provides expertise and financing for exploration and development, but the host government retains total control of petroleum resources. Upon commercial production, the government allocates oil in a manner that allows the contractor to recover its cost from produced oil (cost oil) and allocates royalty and taxes to the government. The remainder of the revenue is shared as profit oil based on the stipulated PSC sharing formula between the host government and the contractor. The extent to which an emergent regime can facilitate a sustainable and responsive administration of resources depends on the corporate and sovereign interests of participating states. Interest of major oil-producing multinational corporations, newer independent producers, and the participating states in the Gulf of Guinea necessitated a critical assessment of the treaty that established the Gulf of Guinea commission. The existing regime must match international best practices for oil and gas exploration and production to manage expectations of all corporate and sovereign stakeholders. The PSC is an essential international framework guiding contractors' and host governments' relationships for sustainability.

Nigeria Deepwater Gas Policy

For the oil industry to continue, countries must replace old reserves through exploration efforts. The

Nigerian deepwater abuts the Sao Tome and Principe deepwater to the extent that a joint resource development program was necessary. A consistent application of the principles of transparency by Nigeria and Sao Tome required more than a report of the amounts of payments made by the IOC to the joint development authority (JDA). Transparency required published drafts of the laws, rules, regulations, guidelines, and standards to administer the joint development zone (JDZ) before they became effective. The two authorities conduct public hearings to elicit opinions from industry and other stakeholders on proposals; publish minutes of meetings, including copies of any reports or other submissions by those participating in the meetings (e.g., consultants hired to conduct due diligence on firms submitting bids); issue reports reflecting the results of the JDA's deliberations, including its decisions to incorporate or reject suggested changes in its proposed standards; and publish the reports of the licensing round committee, including the basis for awarding interests in JDZ acreage, at the time of the announcement of the results of each licensing round (Groves, 2005).

Transparency and free circulation of information eases deficits of weak institutions because civil society, the press, and responsible elements of government use such information to demand accountability and reform. Transparency cannot assure the responsible administration of public resources, but abuse is almost certain without transparency. Oil has been the focus of the Nigerian petroleum industry. The legal and fiscal framework addresses oil production with little or no focus on gas. This is one of the reasons gas exploitation was only a fraction of the capacity utilization for Nigeria. According to Orji (2014), government regimes unsuccessfully addressed gas flaring, gas utilization, and re-injection due to the absence of attractive economic incentives for oil-producing companies to invest in gas utilization or re-injection facilities. Lack of political goodwill and organizational laxity also contributed to prolonged environmental malfeasance from gas flaring (a significant public concern). Gas flaring and conservation reform in Nigeria require the prohibition of flaring or the imposition of flaring fines and attractive fiscal incentives that encourage oil companies to develop gas re-injection or utilisation facilities. The prospect of using litigation to discourage gas flaring is limited by the state's complicity, violating fundamental human rights, and the independence of the judiciary.

Litigation may not facilitate flare reduction or increase investments in gas re-injection facilities. Oil companies delay or challenge anti-flaring suits on technical grounds and appeal decisions that increase the present cost of oil field operations even if existing oil field practices encourage a wasteful pattern of oil production (Orji, 2014). When initial litigation is successful, the process of appeals leads to the Supreme Court. This process takes several years and may not guarantee success for flare reduction or gas re-injection. The most viable approach is to encourage investments in the gas sector through the creation of a secure business environment. This may encourage private sector investments in gas re-injection or utilisation projects via attractive fiscal incentives that encourage oil companies to develop gas re-injection facilities and ensure gas producers can market gas at market prices.

Along with these incentives, the government should increase fines for gas flaring to be commensurate with the value of the flared gas or higher while granting tax incentives to oil-producing companies that establish viable gas utilisation or re-injection options to reduce flares at oil fields. This approach balances attractive fiscal incentives with punitive sanctions. Such policy promotes intergenerational sustainability while balancing the interests of oil-producing companies that worry about the cost of developing gas re-injection or utilisation facilities and the prospects of harvesting profits from such investments due to the existence of government-imposed artificial pricing regimes in the domestic gas market. This approach may be viable for Nigeria as its economy is highly dependent on oil revenue. Any policy that does not substantially accommodate the interests of multinational oil-producing companies may not be enforceable due to the fear that the Nigerian government may lose revenue they require to sustain the fiscal existence of the state.

Nigeria Deepwater Gas Reserves

Nigeria has 187 trillion cubic feet of gas reserves (DPR, 2016). Approximately 52% of this reserve is associated gas and 48% is non-associated gas. The gas accumulation is distributed across the Nigerian terrain of land (30%), swamp (28%), offshore (3%), and deepwater (12%) (DPR, 2016). These natural gas discoveries were incidental to exploration for oil; there was no deliberate effort to explore for gas. Nigeria has copious gas resources and the sector holds huge potentials for growth. The existing legal and regulatory framework, written primarily for oil, does not provide a robust technical and commercial framework for gas. Deliberate policies for the gas sector may provide Nigeria with the opportunity to harness and maximize its stranded gas resources. Effective gas sector development holds great potential to improve the entire Nigerian economy.

CHAPTER 4

COMMERCIALIZATION
OF OIL AND GAS

Generic governmental policies for gas resource commercialization lack specific reference to gas exploitation. According to Ushie, Adeniyi, and Akongwale (2013), fluctuations in oil revenues resulted in inflation, lower output growth, and real exchange rate appreciation in Nigeria. The government hopes to harness the nation's gas resources, mix energy and industrial processes, engage in gas exploration and development to increase the reserve base, encourage indigenous and foreign companies to invest in the industry, and establish infrastructure and incentives to ensure adequate geographical coverage of the gas transmission and distribution network.

According to Ogunleye (2008), Nigeria was rich in crude oil and reaped billions of petrodollars; however, the country struggled to successfully translate oil wealth into sustainable development. Proper management and investment of oil revenue could induce oil-led development in Nigeria if corruption, lack of transparency, accountability, and fairness in use and distribution were no longer problems. Nigeria needs a more pragmatic approach to macroeconomic policy formulation and implementation. Oil revenue volatility harms economic growth but accommodating monetary policy in the form of rising real interest rates often exacerbates this negative effect. Fiscal and monetary macroeconomic policy should align to disconnect the economy from the volatility of oil revenues. The adoption of the medium-term expenditure framework in the budgeting process and benchmarking annual budgets using an oil price-based fiscal rule are steps in the right direction. The enforcement of legislation, particularly the Fiscal Responsibility Act, should underpin the reform agenda. Managing volatility in resource revenues is challenging within a weak institutional context. There is a need for sound oil revenue management institutions, such as a resource wealth fund, to manage the immediate impact of revenue volatility and save revenues for investment in future generations.

Managing oil windfalls in Nigeria required economic policies to reduce volatility and wasteful consumption while channelling oil revenues into productive activities and investments (Ushie et al., 2013). Without a competent bureaucracy to safeguard the interests of the citizenry, implementing policies that will not waste oil revenues is challenging. Mismanagement of oil windfalls in Nigeria was not inevitable. By adopting the right policies and building strong institutions, Nigeria could use its wealth to lift the citizenry out of poverty.

Gas Reserves Commercialization

The main objective of gas reserves commercialization is elimination of flaring associated with oil production. Another objective is the utilization of natural gas for industrial and domestic power generation. In the early 21st century, power generation underwent structural reforms to improve and expand the grid generation capacity and distribution network. Energy security was one of the most urgent public policy issues the Nigerian government addressed to meet its economic development objectives. The goal was to expand the electricity grid and exploit energy sources within Nigeria to ensure a stable supply of electricity.

The government focused on the economic and technical viability of its means to develop energy plans and policies. All independent power plants are now fossil-fuelled systems consisting mainly of gas power. According to Gujba, Mulugetta, and Azapagic (2010), there are concentrated efforts to utilise hydropower even though a major concern exists that the government has not seriously considered environmental and social issues. Renewable energy options (e.g., solar, wind, hydro) do not require burning fuels for electricity generation. Gas is a cleaner fuel than other fossil fuels. Nigeria must attempt to improve the environmental performance of electricity through increased use of renewables; this will require significant trade-offs between economic costs and environmental benefits.

Other important objectives for gas commercialization include the diversification of the economy with gas income. According to Nwoke (2016), changes brought by education, science, and technology created Nigerian entrepreneurs who utilize natural resources that are most prevalent in their environment. The development of indigenous end-user entrepreneurial capabilities, technology acquisition, and diffusion are key element of gas commercialization gains.

Nigeria Gas Stakeholders

There are many stakeholders in the Nigerian gas industry. According to Chan, Watson, and Woodliff (2014), regulators should focus on the corporate governance quality of companies to increase CSR disclosures; there is a correlation between the level of CSR information disclosure and corporate governance ratings that is consistent with stakeholder theory based on firm size, stakeholder power/dispersion, creditor power/leverage, and economic performance. Firms providing more CSR information have better corporate governance ratings, are larger, belong to higher profile industries, and are more highly leveraged.

Since oil discovery in commercial quantities in 1956, different legal regimes governed the sector with weak governance, poor policy implementation, and poor regulation. Cohesion of the stakeholders is essential to effectiveness in the oil and gas industry. According to Idemudia and Ite (2006), corporate-community relationships (CCRs) in the Nigerian oil industry would improve significantly if the needs and aspirations of major stakeholders aligned through a tri-sector partnership. Genuine CSR and CCR initiatives may ameliorate conflicts between host communities and oil companies in the Niger Delta region. Such CSR and CCR initiatives must address the root causes of social conflict. It is therefore imperative that CSR initiatives mitigate the negative effects of existing relationships among key stakeholders.

In the Niger Delta, oil companies made significant efforts to meet their CSR obligations, but were unable to derive maximum benefits from their efforts because CSR initiatives often fail to address the negative effects of oil production or attempt to compensate for the skewed nature of oil exploration. Oil companies often neglect the core issues driving social conflict when designing their CSR agenda. The failure to integrate community perceptions into the design of CSR policies meant that oil companies often experienced conflict with local communities. As a result, oil companies lose the buffer their CSR initiatives provide and are unable to derive maximum benefits.

An expanding range of stakeholders is evolving that encourages businesses of the future to re-invent themselves as forces for good in society. This involves going beyond the paradigm of simply doing no harm and far beyond previous expectations of business being only about shareholders. The abnegation of negative injunction duties by oil companies in the Niger Delta also meant that communities construed affirmative duty obligations (e.g., provision for socio-economic infrastructures) as public relation stunts for oil production companies (Idemudia & Ite, 2006). Community demands for socio-economic infrastructure and development invariably outstrip supply due to the failure of the Nigerian government to meet its fair share of social responsibility in the Niger Delta.

There is growing consensus that the management of social and environmental issues should involve constructive input from three main groups: the industry, government, and civil society. In addition, the private sector in Nigeria must act seriously on CSR and CCR issues and contribute to meaningful development when the government (federal and state level) promotes social, economic, and corporate governance. Governmental efforts must ensure collaboration through restructuring of the oil and gas industry with an all-encompassing modernized PIB.

Gas Utilization in Nigeria

Nigeria has abundant natural gas resources that could provide more energy than the nation's crude oil reserve. There are no specific gas terms in the PSC agreement of 1993 because the focus was always on oil. Lester and Hart (2015) proposed a decentralized strategy for energy technology scale-up, demonstration, and early adoption with greater role for states and regions and a new kind of partnership between the federal government, states, private innovators, and investor. The high costs and risks of demonstrating new clean energy technologies at a commercial scale are major obstacles in the transition to a low-carbon energy economy.

The scaling-up challenges of new technologies are well known. What works in the laboratory or in a small-scale prototype often does not work at full commercial scale. Building, operating, and debugging full-scale prototypes invariably reveals new problems. New technologies rarely deploy in isolation. More often, they begin as part of pre-existing technological and organizational systems; the task of integration is often very demanding (Lester & Hart, 2015). Complementary technologies (e.g., new manufacturing processes and logistical systems) must scale-up in parallel to obtain effective synergy for sustainability. Many gas projects are at various stages of inception and commissioning; this implies avid support by the government for the utilization of gas to improve the economy.

The Nigerian Gas Company (NGC), a subsidiary of NNPC, currently supplies gas for power generation, fuel, or feedstock to cement and fertilizer plants, glass plants, food and beverages companies, and various other manufacturing industries. More local industries are using gas, increasing demand for gas utilization. The Nigerian gas market is a profit-oriented market with attractiveness for investment. Effective gas utilization may benefit from export-oriented projects planned by the NNPC and its JV partners. Such projects include: the Escravos Gas Project by NNPC/Chevron JV, the Oso NGL project by NNPC/Mobil JV, and six trains of the Nigeria LNG projects by the NNPC/Shell/Agip/Total JV exporting LNG with the seventh train in construction. The West African Gas Pipeline (WAGP) project will supply gas to some Economic Community of West African States (ECOWAS) countries, pursuant to the ECOWAS treaty, which encourages member nations to cooperate, consult, and coordinate policies regarding energy and mineral resources. The project has commercial viability and technical feasibility (WAGP Company, 2016).

Nigeria Liquefied Natural Gas (NLNG) incorporated to harness Nigeria's vast natural gas resources and produce LNG and natural gas liquids (NGLs) for export (NLNG, 2016). The shareholders of the company include: the federal government of Nigeria represented by NNPC (49%); Shell (25.6%); Total LNG Nigeria Ltd. (15%), and Eni (10.4%). NLNG's wholly-owned subsidiaries include: Bonny Gas Transport (BGT) Limited and NLNG Ship Management Limited (NSML). One of the aspirations of NLNG was to maintain exceptional standards in community relations and technology transfer to Nigerians to promote sustainable development of Nigerian businesses.

CHAPTER 5

PERSPECTIVES OF KEY STAKEHOLDERS

A key focus group for consideration in the project is community members who understand the impact of gas exploration on their lives. Another key group is oil industry employees who work in gas production. A third important focus group is government personnel who work on policies regarding oil and gas. Gathering perceptions of members of each of these groups may improve dialogue, increase cooperation in defining acceptable solutions, enhance social cohesion, and lead to ownership of decisions and acceptable outcomes for sustainable deepwater gas development. The host communities are at the lowest rung of the ladder of influence followed by the oil companies; the government is at the top of the ladder of influence.

Niger Delta Community Issues

Livesey (2001) explored socio-political conflicts that reflect the dynamics of cultural and institutional change by multinational companies operating in Niger Delta. The wealth extracted from the ground did not translate into monetary and developmental benefits for the region, and the Niger Delta remains one of the most impoverished regions in the world. Environmental degradation due to the exploitation of oil over the years exposed the people to environmental hazards that compounded health problems, a lack of safe water, devastation of arable farmlands, and destruction of the fishing industry. There is a connection between gas flaring, health problems in the communities, and poor agricultural yields.

Economic benefits, political allegiance, and religious views all compete with education, awareness, and experience of gas flaring in the community; some models of education, awareness, and experience distort perceptions of environmental problems in communities where political tension and economic adversity are prevalent (Edino et al., 2010). Years of broken promises by multinational oil companies under the guise of non-interference into government's legitimate duties created poor relationships between oil companies and communities. Organized protests and restiveness developed to address the grievances as citizens campaigned for their rights to control resources from their land. The government's reaction to organized protest through its security officials was often excessive and repressive.

According to Ordinioha and Brisibe (2013), oil spills contaminated the surface water, ground water, ambient air, and crops with hydrocarbons including known carcinogens like polycyclic aromatic hydrocarbon and benzo(a)pyrene, naturally occurring radioactive materials, and trace metals that further bioaccumulated in some food crops. Oil spills are typically caused by poor maintenance of aging facilities and vandalizing of pipelines by oil thefts. The mangrove forests of the Niger Delta are an important ecological resource that provides an essential ecosystem for soil stability, medicines, healthy fisheries, wood for fuel and shelter, tannins and dyes, and critical wildlife habitats. Oil companies realize the importance of pollution prevention and routinely incorporate environmental impact assessments into their corporate strategies (Eweje, 2006).

The people of the Niger Delta reacted violently to frustrations when the Nigerian government failed to satisfy their socio-economic needs. The frustration also resulted from the devastation of the environment of the region, pervasive poverty and underdevelopment, legislative disempowerment and subjugation, inability to control the crude oil resource, and suppression of the people by the state. The oil company's CSR, as an anti-conflict strategy for development, was an obligatory framework to separate social investment from operational costs (Ojo, 2012). According to Ezeani (2012), Nigeria experienced significant challenges in constructing a stable legal framework for trade and development. Economic policy-making in Nigeria is often a rapid-fire response to problems through legislation that is largely speculative to cover all eventualities. The government considers critical areas of action to be those involving punitive measures that its agency undertook. These constraints are challenges for Nigeria's legislature, which bears the burden of creating laws that can address trade and development needs.

Direct contributions of the organized private sector (OPS) and the involvement of various professionals in different trades and industries are essential. The Manufacturers Association of Nigeria; Nigeria's Association of Chambers of Commerce, Industry, Mines and Agriculture; and the Nigerian Employer's Consultative Association are the foremost participants in the OPS. However, their contributions are still subordinate to institutional initiatives led by the government. Nigeria needs the independent contributions of different constituents of the OPS, the institutional responses of government bodies such as the Ministry of Commerce, and the skills of persons learned in international economics and international economic law (e.g., trade negotiators and policy makers) to inform policies.

The objective of effective policy-making is not, as has been the case so far, the adoption of many proposed reforms by each subsequent political administration, but the adoption and implementation of crucial objectives that visibly impact the socio-economic environment in the country. There is an absence of definite jurisprudence in Nigerian government policy. This demonstrates a lack of progress of public law and judicial review of administrative action. According to Ezeani (2012), the activities of the government regarding policymaking and implementation are hardly ever subject to judicial determination.

Government Policy Makers

Public interest was paramount in the minds of government policy makers ensuring maximum benefit from the oil wealth of the nation. Implementation of the extractive industries transparency initiative (EITI) promoted accountability and effective management of resource revenues. The challenge facing Nigeria was how to convert the revenue into a tool for lifting the present generation of Nigerians out of poverty. According to Nwapi (2014), the challenge was to debunk the pessimistic view that with rising resource revenues, the quality of governance declines, impoverishing the population and making the poor people poorer. Nigeria made great advances in applying the principles of EITI to policy formulation, yet an incongruity exists between achievements through the policy and the socio-economic situation in the country.

A number of factors are responsible for the situation: inability to enforce accountability measures; poor quality of institutions; improper application of rule of law and property rights; preponderance of vested political interests; inconclusive investigation of corruption cases; lack of fiscal and development coordination between the national and the subnational governments; uneven development across the country;

and complicated oil and gas contract systems with intrinsic difficulty in monitoring the oil and gas industry. Improvements in public welfare and better developmental outcomes, more equitable distribution of wealth, improved socio-economic conditions, and poverty alleviation are potential benefits of improved policy. The government adopted CSR policies to support domestic political-economic institutions and international competitiveness of domestic businesses (Knudsen & Brown, 2015).

As businesses became increasingly global and began sourcing from and producing in developing countries without adequate social and environmental protection, companies became more interested in government support. The competitive position of firms informs policy structures. The ability of businesses to influence governments and alter policy is contingent on relations between businesses and governments including the economic importance of specific firms. The Nigerian government supported CSR through multinational oil companies. For example, a policy required 2% of profit by oil companies be set aside for educational development of the oil producing communities. This was a key poverty alleviation strategy because over 70% of the population in oil producing communities lived below poverty line (Knudsen & Brown, 2015).

Non-Government Organizations

Rising violence in the Niger Delta resulted from criminal tendencies due to the insensitivity of the government. An imbalance of power caused violent acts and deep-rooted destructive upheaval in Ogoni land where Shell first discovered oil in commercial quantities in the Niger Delta. Catastrophic failure in relations with the Ogoni and the consequent fall-out with non-government organizations (NGOs) prompted Shell to review its business principles and initiate a multi-million-dollar exercise in stakeholder outreach and communication (Wheeler, Rechtman, Fabig, & Boele, 2001). However, lack of trust and reconciliation of stakeholder priorities resulted in suboptimal delivery and execution of the community-based projects.

The portrayal of Shell as a foreign oil giant with an unimpressive record of social and environmental concern is a reputational issue for the company. The armed conflict between militias and government forces in the Niger Delta spanned over two decades, defying all solutions. A disarmament, demobilization, and reintegration (DDR) program began in August 2015 to end the violence and remained in place. It was a radically different approach that displayed zero tolerance for political challenges to oil production or the allocation of oil profits. The DDR program forced a ceasefire, engaged militants in planned programs to rehabilitate and reintegrate them into civilian society, and opened the oil wells, many of which closed due to the crisis (Okonofua, 2016).

This broad policy shift created a new set of issues requiring critical assessment of the Niger Delta Amnesty Program, which had implications for adaptation and implementation of the DDR program. The Niger Delta conflict transformed into intense contestations of provisions to deprioritized women in the communities. The Amnesty program in the Niger Delta was a form of pacification rather than a resolution of fundamental issues. Bunkering, piracy, and kidnappings resulted. Unprecedented malfeasance and a lack of effective means to address ecological tragedies from oil spills and gas flaring continued. Sabotage-related oil theft resulted from youth unemployment. Compensatory gestures by

Shell only fed into institutionalized corruption without adequately addressing issues of public concern.

CHAPTER 6

POTENTIAL BENEFITS OF

DEEPWATER GAS RESERVES

The deepwater gas reserves in Niger Delta hold significant potential contributions to employment, technological development, government revenues at different tiers and for different stakeholders, lease acquisition and royalties from production, and both domestic and international contribution to Nigerian energy needs. Stakeholder analysis yielded three key stakeholders in the Niger Delta: community, government, and oil industry players. The management of social and environmental issues should involve constructive input from these three main groups. Understanding of the deepwater gas value chain can be deduced from the utilization of gas from onshore and the production of associated gas from deepwater oilfields. The Nigerian government's cautious approach to the regulation of technically challenging deepwater oil and gas exploration resulted in a PSC exclusively for oil extraction without favorable terms for deepwater gas extraction. The Niger Delta turned out to be an essentially gas province with lots of opportunities for gas exploitation. Current deepwater gas extraction is limited to the unavoidable production of associated gas alongside oil production since there is currently no working gas agreement to extract non-associated gas reserves. Oil industries and government regulating agencies are organized institutions that play significant roles in the integration of resource extraction and its impact on people and the environment.

There has been no deliberate search for gas in the deepwater because there are currently no working gas terms in place. Government policy for the profitable extraction of oil and gas has been selectively executed by multinationals (i.e., business people with overarching business interests). The lack of local technical capacity for oil and gas exploitation as well as politically induced institutionalized corruption resulted in value erosion for the community from oil and gas natural resources. Expected capital investment in the extraction of deepwater gas is enormous; yet, sizeable past discoveries and new potential finds must result in a deliberate search effort to provide commensurate return on investment. The creation of a business environment with adequate incentives will encourage investments in gas utilization, which should be balanced with punitive sanctions for gas flaring.

Mining Lease and Royalties

The Nigerian government found the PSC funding arrangement for recovering investment attractive; however, violence and insurgency plagued the oil and gas revenue sharing system at all levels of governance. The opening up of the deepwater acreages for gas exploration and exploitation will no doubt attract a flurry of international investors because of the comparatively cleaner and more environmentally friendly utility for gas compared with oil. Crashes in global oil prices shook Nigeria's confidence in meeting obligations as a nation due to overdependence on oil resource revenues. The resource curse also affected Nigeria due to a lack of public accountability, excessive external debt, and a lack of diversification to other sectors. The relative stability of gas in the international market compared to oil price is another advantage for deepwater gas extraction. The federal government typically receives significant revenues in the form of signature bonuses during lease sales. The IOCs liked the 1993 PSC terms; therefore, the federal government is keen to make amendments for optimum national benefit in the event of deepwater gas exploration and exploitation agreements. The government revenues from royalties and possible stakes in investment for deepwater gas extraction will be substantial.

Employment

The development of deepwater gas resources is expected to yield noteworthy employment opportunities in the Niger Delta states and nationally. The regulation already in place to ensure local content in the development of oil and gas projects will no doubt augur well for the job market in Nigeria. The youth population will benefit from the opportunity for gainful employment in the oil and gas industry stemming from deepwater natural gas exploration and production.

According to the National Bureau of Statistics, the youth employment rate as of 2016 was 14%. The impact on the economy will be widespread among ancillary industries and service companies including: oil and gas exploration, development, and production companies; oil and gas technical services such as engineering; manufacturing and fabrication of oil and gas equipment and platforms; professionals in different cadres; construction companies; and scientific research companies. Healthcare delivery and social assistance will increase as well as food services, administrative support, waste management services, finance and insurance sectors, real estate leasing, and other rental services.

Oil companies typically employ about 40% technical personnel and 60% service support personnel due to the complexity of the oil and gas extraction business. The quest of the Nigerian government to actively participate in oil and gas business in partnership with IOCs for possible technological transfer further exacerbated the complex relationship of the government and IOCs. There may be job opportunities in government sectors as well after beginning deepwater gas extraction. An optimum balance between regulatory oversight of some government agencies and active participation as majority shareholding partner by another government agency is difficult to achieve.

Development Projects and Drilling

Deepwater gas development projects would provide key facilitation in natural gas production. These projects will be capital investments and operational expenditures culminating in employment, economic activities, and governmental revenues. The complexities of offshore gas projects elicit the best contributions from top-notch engineers, contractors, and various service providers for several years prior to production and transport of gas to the markets.

Dedicated drilling activity to explore for gas will be an important activity by the oil and gas companies and contractors. Despite probable sizeable accumulations of non-associated gas, there has never been a deliberate search for deepwater gas. Drilling activities will continue during appraisal stages to ascertain the extent of accumulation with attendant economic and commercial valuation of the finds. Further drilling activity will occur during the gas development and production stages and dedicated gas production wells will be drilled. Valuable technological application of world-class safety measures in such deepwater gas ventures are expected as part of the technology transfer to Nigerians.

Natural Gas Production and Revenues

The development of deepwater gas could lead to an increase in quantity of gas available for both domestic consumption and international supply. The deepwater portion of Nigerian gas is currently considered significant without any potential finds that may result from deliberate searches for gas. Several deepwater gas reservoirs are close together, providing an opportunity for a limited environmental footprint during production of the deepwater gas accumulations in the Niger Delta. Increased revenues from the additional production from deepwater gas may result in more money for all key stakeholders.

Initial investments in deepwater gas extraction may be significant, but they will come from the IOCs if the extraction agreements are an improvement over the deepwater oil extraction terms. Different categories of activities related to finding, developing, and producing deepwater gas require different amounts of capital. Some activities have more substantial impact on overall spending (e.g., drilling of gas wells, operations, engineering, equipment manufacturing, and fabrication of production platforms). There are equally complex supply chains in developing deepwater gas that will open activities in specialized equipment component manufacturing far from areas of exploration and production. A consensus solution is necessary to assure equitable representation of interests of all key stakeholders in the business.

Expenditure on the deepwater gas development project and the impact of increased revenues to various tiers of governments should result in a remarkable increase to Nigeria's GDP. Gas deposits are significantly larger than oil deposits in the Niger Delta. Cumulative revenue from royalties, taxes, and production sharing agreements could alleviate poverty in the Niger Delta area depending on the federal government's wealth sharing formula. Although the positive effect of deepwater gas production will be felt nationwide, most of the job creation, economic, and revenue impact of increased non-associated gas production will occur in the coastal states of the Niger Delta.

There will be an increase in Nigeria's energy security due to the advantage of available production from deepwater gas projects. Fluctuation in international gas prices is less than that of oil. A more equitable balancing of the energy needs of the nation could occur by pursuing deepwater gas production. The quest to ensure a stable supply of electricity led to public policy issues on economic development. Steel, fertilizer, and petrochemical industries use gas as a clean fuel alternative. The government may be able to derive the bulk of necessary gas from deepwater when monetization policy issues are resolved.

CHAPTER 7
CONSENSUS
SOLUTIONS

According to Lewis (2001), a stakeholder is any person or group that can make a claim on an organization's attention, resources, or output or who the organization may affect. Stakeholder theory conceptualizes the social, political, economic, communal, and racial identities of people who serve a common interest (Wagner, Alves, & Raposo, 2011). The present research included participants from the communities of Port Harcourt, Warri, and Yenagoa, government regulatory agencies, and the oil industry. There are conflicts of interest between these stakeholders. According to Jensen and Sandstrom (2011), globalization brings new power relations and new dimensions of responsibility to stakeholder theory. The multinational oil companies should take responsibility for profitable and safe extraction of deepwater gas; the government must ensure a favorable pricing regime, and the community must cooperate. A fair approach is more effective for attracting, retaining, and motivating reciprocal stakeholders to create value (Bridoux & Stoelhorst, 2014).

Public and private institutions operate according to external factors (e.g., environment) and internal factors. Institutional theory explores how people construct boundaries between businesses and society. Application of institutional theory may improve understandings of the effectiveness of CSR within the wider field of economic governance (Brammer & Jackson, 2012). Tensions between business-driven and multi-stakeholder forms of CSR extend to the transnational level where the form and meaning of CSR remain highly contested. Institutional theory supports the development of organizational structures of control that reflect the expectations of society. Oil companies and government regulatory agencies must be responsible private and public institutions that satisfy the social and economic needs of their communities of service. According to Garrouste and Brousseau (2011), institutions impose rules on individuals for strategic coalition, but also seek some form of equilibrium. Oil companies pay all taxes and royalties the government imposes, and government agencies monitor compliance and grant statutory approvals. The focus group responses included discussion of the effective fulfilment of these roles.

The participants completed a research questionnaire (see Appendix A) to: (a) establish knowledge about oil and gas resource and income generation; (b) establish knowledge about government policy to exploit and monetize deepwater gas for profit; (c) determine the possibility of utilizing deepwater gas in the same way as onshore gas for domestic purposes; (d) determine the government's interest in giving incentives through monetization policies to help harness the deepwater gas profitably; (e) determine if monetization of deepwater gas could help industrialization by empowering the steel, fertilizer, and petrochemical industries; (f) determine the effectiveness of the focus group members in influencing government decisions on deepwater gas monetization policies; (g) determine the government benefit of global experience of multinational oil companies in shaping deepwater gas pricing; and (h) determine perceived benefit to communities from harnessing deepwater gas deposits.

According to Swanson et al. (2010), adaptive techniques function effectively in complex, dynamic, and uncertain conditions, such as the conditions in communities of the Niger Delta region. The researcher conducted focus groups (see Appendix B) and analyzed responses with respect to perspectives on a proposed policy for gas monetization. According to Curry (2013), growth in polices and plans increases community participation. The researcher included local communities in this study despite their low level of influence in the stakeholder influence diagram and included interviews with senior and mid-level management staff of government regulatory agencies.

Participants differed in gender, age, national origin, ethnicity, disability, sexual orientation, education, and religion, which reflects the population of most 21st century organizations in an increasingly global market (Cao, Clarke, & Lehney, 2003). The researcher conducted narrative analysis of the survey responses to compare opinions and perceptions that reflected true diversity in the views of the different respondents in the stakeholder groups. Community members from Delta, Rivers, and Bayelsa states were affiliated with grassroots movements and continue to live with the consequences of oil and gas exploitation in their homelands. Participants in the community stakeholder focus group included community representatives who lived in target communities for at least 10 years.

The oil company focus group included business people seeking profitable investment in oil and gas ventures. One of the foremost multinational oil and gas companies operating in Nigeria since the 1930s represented the oil and gas companies. They have a responsibility to safely extract oil and gas without environmental malfeasance. The researcher included employees from this company to represent the oil industry in the focus group of stakeholders.

The government created policies that regulated oil and gas extraction by oil companies. The researcher included two main government agencies in the government focus group for discussion of deepwater gas monetization. The inclusion of three distinct focus groups with pivotal stakeholder groups was essential to fully explore perceptions of deepwater gas monetization. In all, 27 people provided robust questionnaire responses and focus group discussions. The diverse spectrum of responses provided richer data for analysis.

Oil company employees included line staff, team leaders, and managers who lived in one of the three operational locations of the company in Nigeria. Government participants included the government regulatory agency's line staff and managers. Members of the community included individuals who lived in Port Harcourt, Warri, and Yenagoa who had ties to grassroots organizations. The participants were between the ages of 25 to 60 years old, male and female, and varied in occupation (e.g., artisans, professionals) and economic status. The researcher obtained permission from the DPR to engage oil industry employee in gathering data, a statutory requirement. For the community group, participation was limited to community members living in cities due to logistics.

Surveys and Focus Groups

The questionnaire was a survey of individual opinions whereas focus group discussions formed team responses. Analyzing the data yielded an in-depth understanding of participants' dispositions about policies. Perceptions of all stakeholders regarding these issues were an essential part of the study to gain a critical understanding of community issues. The researcher conducted a qualitative comparative analysis to inform a possible intervention (Blackman, Wistow, & Byrne, 2013). The researcher contextualized comments from group discussions according to responses from the questionnaire to corroborate or refute the statements. Multiple presentations of data provided various means of comparing data (e.g., charts, posters, photographs, booklets, and summary reports). Technical information in the oil and gas industry appears in this manner for clarity and ease of understanding by consumers. The researcher gathered qualitative data from different focus group members, analysed, and synthesized the data to discover meaningful conclusions for the oil and gas industry and Nigerian government. Consensus among focus groups and questionnaire survey data may be a potent tool for an informed policy initiative on the deepwater gas monetization.

Findings Reflect Similarities in Disparate Groups

The research revealed an interconnectedness of various issues concerning oil and gas extraction in the Niger Delta through the process of qualitative data acquisition and analysis. A narrative analysis of the data revealed implications of respondents' perceptions related to extraction of deepwater gas in the Niger Delta. The main data sources were the individual survey responses and the focus group thematic narratives.

Individual Survey

The type of data generated from the survey included specific yes/no answers as well as graduated scale responses with specific ranges. Tables 1, 2, and 3 show the survey responses from each stakeholder group.

Table 1

Survey Responses by Community Participants

Questions	Answers		
1a. Do you have any affiliation to the oil and gas industry in Nigeria either by work, place of abode or peculiar interest?	Yes	No	
	10	0	
1b. Are you willing to share your perspectives about deepwater gas deposits in Niger Delta without any bias?	Yes	No	
	10	0	
1c. Age	31-40	41-50	51-60
	7	3	0
1d. Gender	Male	Female	
	9	1	
1e. Education	High School	Graduate	Post graduate
	0	8	2
2a. About what percentage of Nigeria's foreign exchange earning comes from oil and gas?	26-50%	51-75%	>75%
	0	2	8
2b. About how much gas reserves does Nigeria have?	100-200tcf	200-300tcf	>300tcf
	2	5	3
2c. About what percentage of the gas is located in deepwater Niger Delta?	<1/3	1/3 - 2/3	>2/3
	0	8	2

Questions	Answers		
3. Is there a government policy to exploit the deepwater gas for profit?	Yes 5	No 2	Don't know 3
4. Government has the Nigerian Liquefied Natural Gas (NLNG) trains that have been exporting gas for profit. Do you think government can profitably monetize the deepwater gas for export at same pricing?	Yes 9	No 0	Don't know 1
5. Government (in affiliation with her partners in the joint venture) has been utilizing gas from onshore for domestic uses. Do you think the deepwater gas can be utilized in a similar manner?	Yes 8	No 2	Don't know 0
6. Gas from deepwater is typically more expensive to harness than onshore. Do you think the Nigerian government can give adequate incentives through monetization policy to help harness the gas profitably?	Yes 10	No 0	Don't know 0

Questions	Answers		
7. Do you see any connection between gas utilization from deepwater and industrialization?	Yes 10	No 0	Don't know 0
8. Do you think the deepwater gas can help power the steel industry, the fertilizer and petrochemical plants?	Yes 10	No 0	Don't know 0
9. Do you think Nigerian government will be willing to grant special incentives for deepwater gas monetization to support growth and industrialization?	Yes 8	No 0	Don't know 2
10. Do you think this kind of research with opinions from oil industry, government and community personalities could help government decide her monetization policies?	Yes 10	No 0	Don't know 0

Questions	Answers		
11. Do you think government can benefit from global experience of multinational oil companies in determining deepwater gas pricing?	Yes	No	Don't know
	10	0	0
12. Do you think there is real benefit to the communities from harnessing the deepwater gas deposits?	Yes	No	Don't know
	9	1	0
13. In your opinion, which of the three groups below will have the most benefit from harnessing the deepwater gas deposits?	Multinational oil companies	Gov. (fed./ state/ local)	Communiti es in the Delta
	2	8	0
14. In your opinion, which of the three groups below will have the least benefit from harnessing the deepwater gas deposits?	Multinational oil companies	Gov. (fed./ state/ local)	Communiti es in the Delta
	1	0	9

Table 2

Survey Responses by Oil Industry Participants

Questions	Answers		
1a. Do you have any affiliation to the oil and gas industry in	Yes	No	
Nigeria either by work, place of abode or peculiar interest?	10	0	
1b. Are you willing to share your perspectives about deepwater gas	Yes	No	
deposits in Niger Delta without any bias?	10	0	
1c. Age	31-40	41-50	51-60
	2	4	4
1d. Gender	Male	Female	
	9	1	
1e. Education	High School	Graduate	Post graduate
	0	4	6
2a. About what percentage of Nigeria's foreign exchange	26-50%	51-75%	>75%
earning comes from oil and gas?	0	1	9
2b. About how much gas reserves does Nigeria have?	100-200tcf	200-300tcf	>300tcf
	1	3	6
2c. About what percentage of the gas is located in deepwater Niger	<1/3	1/3 - 2/3	>2/3
Delta?	2	5	3

	Yes	No	Don't know
3. Is there a government policy to exploit the deepwater gas for profit?	0	6	4
4. Government has the Nigerian Liquefied Natural Gas (NLNG) trains that have been exporting gas for profit. Do you think government can profitably monetize the deepwater gas for export at same pricing?	3	6	1
5. Government (in affiliation with her partners in the joint venture) has been utilizing gas from onshore for domestic uses. Do you think the deepwater gas can be utilized in a similar manner?	8	1	1
6. Gas from deepwater is typically more expensive to harness than onshore. Do you think the Nigerian government can give adequate incentives through monetization policy to help harness the gas profitably?	9	0	1
7. Do you see any connection between gas utilization from deepwater and industrialization?	8	2	0
8. Do you think the deepwater gas can help power the steel industry, the fertilizer and petrochemical plants?	9	0	1

9. Do you think Nigerian government will be willing to grant special incentives for deepwater gas monetization to support growth and industrialization?	Yes	No	Don't know
	8	0	2

10. Do you think this kind of research with opinions from oil industry, government and community personalities could help government decide her monetization policies?	Yes	No	Don't know
	8	0	2

11. Do you think government can benefit from global experience of multinational oil companies in determining deepwater gas pricing?	Yes	No	Don't know
	10	0	0

12. Do you think there is real benefit to the communities from harnessing the deepwater gas deposits?	Yes	No	Don't know
	7	1	2

13. In your opinion, which of the three groups below will have the most benefit from harnessing the deepwater gas deposits?	Multinational oil companies	Gov. (fed./state/local)	Communities in the Delta
	1	9	0

14. In your opinion, which of the three groups below will have the least benefit from harnessing the deepwater gas deposits?	Multinational oil companies	Gov. (fed./state/local)	Communities in the Delta
	1	0	9

Table 3

Survey Responses by Government Participants

Questions	Answers		
1a. Do you have any affiliation to the oil and gas industry in Nigeria either by work, place of abode or peculiar interest?	Yes	No	
	6	1	
1b. Are you willing to share your perspectives about deepwater gas deposits in Niger Delta without any bias?	Yes	No	
	6	1	
1c. Age	31-40	41-50	51-60
	2	3	2
1d. Gender	Male	Female	
	7	0	
1e. Education	High School	Graduate	Post graduate
	0	3	4
2a. About what percentage of Nigeria's foreign exchange earning comes from oil and gas?	26-50%	51-75%	>75%
	0	0	7
2b. About how much gas reserves does Nigeria have?	100-200tcf	200-300tcf	>300tcf
	5	1	1
2c. About what percentage of the gas is located in Niger Delta?	<1/3	1/3 - 2/3	>2/3
	0	7	0

3. Is there a government policy to exploit the deepwater gas for profit?	Yes	No	Don't Know
	5	1	1

4. Government has the Nigerian Liquefied Natural Gas (NLNG) trains that have been exporting gas for profit. Do you think government can profitably monetize the deepwater gas for export at same pricing?	Yes	No	Don't know
	5	2	0

5. Government (in affiliation with her partners in the joint venture) has been utilizing gas from onshore for domestic uses. Do you think the deepwater gas can be utilized in a similar manner?	Yes	No	Don't know
	5	1	1

6. Gas from deepwater is typically more expensive to harness than onshore. Do you think the Nigerian government can give adequate incentives through monetization policy to help harness the gas profitably?	Yes	No	Don't know
	7	0	0

	Yes	No	Don't know
7. Do you see any connection between gas utilization from deepwater and industrialization?	7	0	0
8. Do you think the deepwater gas can help power the steel industry, the fertilizer and petrochemical plants?	6	1	0
9. Do you think Nigerian government will be willing to grant special incentives for deepwater gas monetization to support growth and industrialization?	6	1	0
10. Do you think this kind of research with opinions from oil industry, government and community personalities could help government decide her monetization policies?	5	1	1
11. Do you think government can benefit from global experience of multinational oil companies in determining deepwater gas pricing?	7	0	0

	Yes	No	Don't know
12. Do you think there is real benefit to the communities from harnessing the deepwater gas deposits?	6	1	0

	Multinational oil companies	Govt. (fed./state/local)	Communities in the Delta
13. In your opinion, which of the three groups below will have the most benefit from harnessing the deepwater gas deposits?	1	6	0
14. In your opinion, which of the three groups below will have the least benefit from harnessing the deepwater gas deposits?	0	0	7

Focus Groups

The data from focus group discussions with the three different stakeholder groups were qualitative comments about their opinions. The data from the three focus groups appear as narrative summaries in the following sections. The researcher considered agreement from all participants on an issue a consensus and considered agreement of 80% of all participants a majority. These analyses enabled identification of issues of key concern (consensus by all key stakeholders) that may influence recommended actions for relevant stakeholders.

Community stakeholder group. All respondents in the community stakeholder group answered yes to their affiliation with the oil and gas industry in Nigeria through work or place of abode. They all confirmed interest in the relationship of the community with the oil and gas business. All respondents were willing to share their perspectives and opinions about deepwater gas deposits in Niger Delta without any bias. Seven of the respondents were age 31 to 40 and the remaining three were between 41 and 50 years old. The age ranges ensured all respondents met the ten-year residency requirement in the Niger Delta community as an adult. There were nine male respondents and one female respondent; all had graduate degrees from institution of higher learning and two respondents had post-graduate degrees.

Eight of the respondents agreed that over 75% of Nigeria's foreign exchange earnings came from oil and gas; two respondents estimated it was between 50 and 75%. Two respondents thought that Nigeria gas reserves were between 100 and 200 trillion cubic feet; five others thought they were between 200 and 300 trillion cubic feet. The last three thought they were over 300 trillion cubic feet. On the split of gas deposit between onshore and deepwater, eight respondents thought that one-third to two-thirds existed in deepwater; two respondents thought it was greater than two-thirds.

Five respondents said government policy existed to exploit deepwater gas for profit; two respondents disagreed. Three respondents did not have an opinion regarding government policy on exploiting deepwater gas. Nine respondents thought the Nigerian government can profitably monetize deepwater gas for export at the same pricing as the NLNG; one respondent did not have an opinion. Eight respondents thought the government could harness deepwater gas for domestic use similar to onshore gas; two respondents disagreed. All respondents thought that the government could give adequate incentives through monetization policy to improve deepwater gas profitably. All respondents expressed a positive connection between utilization of deepwater gas and industrialization. Eight respondents thought the Nigerian government would be willing to grant special incentives for deepwater gas monetization to support growth and industrialization, and two respondents had no opinion. All respondents thought this study could help the government improve monetization policies, and that the government could benefit from the global experiences of multinational oil companies in determining deepwater pricing.

Nine respondents thought harnessing deepwater gas deposits would benefit the community; one respondent disagreed. Eight respondents thought that the government would benefit most from deepwater gas deposits; two respondents thought multinational oil companies would benefit most. Nine respondents thought local communities would have the least benefit from harnessing deepwater gas; one respondent thought multinational oil companies would have the least benefit.

The group suggested that the capacity existed for infrastructure development, but policy lacked sustainable management. Focus group participants shared that some remote villages use flared gas as a source of light, which is detrimental to the environment. A participant affirmed that the "government lacks the will to convert flared gas to electricity." "Flared gas was also used to dry cassava by some of the local women in the communities," according to another participant.

The effective enforcement of policies already in place with requisite accountability from government agencies could improve competitive pricing of the deepwater gas reserves. However, government policy is inconsistent, unfair, and slow. Participants noted infrastructure decay and little or no capacity for drilling in deepwater. The government was not willing to invest in provision of electricity for community areas and made little or no monetary profit from gas being flared. There was perceived neglect of community issues in government policies on provision of infrastructures in remote communities. A participant commented "government in not forthcoming," implying a lackadaisical attitude of government towards their plight as community dwellers.

Community members described the presence of multinational oil companies as both positive and negative. Multinationals were willing to harness deepwater gas if policies and incentives that considered all environmental and security issues were well articulated. According to a participant, "multinationals will harness gas if incentives are right." This lends credence to the idea that beneficial government policy could attract multinationals to help unlock the economic value of deepwater gas. Participants noted that oil and gas exploration and production significantly affected the lives and well-being of people all over the world. Multinationals were venturers in the lucrative oil business that would exploit policy loopholes for monetary benefit.

Participants explained the negative impact of oil and gas exploitation in the community: oil spills, gas flares, effluent discharges, contamination, biodiversity impact, and livelihood eradication. A participant said "environmental impact is taking its toll on the community." Food shortage, groundwater contamination, and attendant health issues were themes related to the operation of multinational oil and gas companies. Participants explained their belief that multinationals dispose of onshore assets and move towards deepwater to reduce their environmental footprint and interaction with communities. The clash between host communities and oil companies is an ongoing issue that would continue even with operations in deepwater.

Analysis of community discussion. Dwelling on the treatment of host communities in the process of oil and gas extraction, some community participants opined that IOCs carry out CSR beyond expected levels of performance due to the government's inability to provide adequate security and favorable policy.

Participant comments like "the oil companies are trying their best" is indicative of the efforts some oil companies made to ameliorate the disturbing consequences of oil and gas extraction in the communities. Destructive activities (e.g., stealing crude oil from pipelines) added to environmental degradation in the communities. Participant comments such as "there is big clash between host communities and oil companies" is indicative of the level of disaffection between some of the key stakeholders. Corruption existed; community representatives diverted money from multinationals for community projects to personal interests.

Participants supported utilization of deepwater gas for electricity generation. Gas provides cleaner energy and participants noted it was a springboard for economic growth in Nigeria. They also noted that the creation of a market (either internal or external) was the most important first step in realizing the potential of deepwater gas. The enforcement of workable policy through grassroots participation may benefit deepwater resource exploitation. Proper monitoring of viable policy would assure a re-investment of the proceeds from gas into infrastructure development. Building upon earlier legacies by successive governments could ensure continuity of deepwater gas development programs.

Participants emphasized the importance of the environment and sustainable use of deepwater gas. One participant stated, "gas provides cleaner energy and thereby has prominent role in the future." The move of the world towards cleaner, more environmentally friendly energy source makes deepwater gas attractive. Table 4 includes discussion responses by the community focus group members.

Table 4

Discussion Responses by Community Focus Group

Question A	What is your opinion about government policy on gas deposits in the Niger Delta?
Positive Comments	Capacity for infrastructure development. Resource base is available. Proper policy management will result in infrastructure development. Recent West African Gas Pipeline is a good development. Flared gas is source of light in remote villages.
Neutral	Competitive pricing is very important. How do we make government responsible to account for policy implementation? How effective are the policies already in place?
Negative Comments	Government is not forthcoming. Inconsistent policy. Government policy not fair. Government policy is slow. Infrastructure decay. No capacity to drill in deepwater. Losing so much gas to flaring (about 1.2bcf/d). Gas market is a big challenge. No electricity. Government lacks the will to convert flared gas to electricity. Government regulatory requirement and policies does not cover effectively the community issues.
Question B	What do you think of the big oil companies' presence in deepwater Nigeria?
Positive Comments	Multinationals will harness gas if incentives are right. A policy that takes into account all environmental and security impact is essential.
Neutral	Multinationals have big presence in deepwater. Their presence could be positive or negative. Oil and gas explorations

	determines the wellbeing of the people all over the world.

Oil is more lucrative for the multinationals. Multinationals could exploit loopholes in our policy. Environmental impact is taking its toll on the community. Oil spills, gas flares, effluent discharge, contamination, biodiversity impact, and livelihood eradication. Food shortage due to exploration and production activities of oil and gas companies. Destruction of zinc roof, underground water contamination and health issues introduced by oil and gas company operations. Deepwater has less responsibilities to the community. Gas contaminates the water in the community. Multinationals are selling onshore assets and moving to offshore to reduce environmental footprints and interaction with communities. Some services sourced outside such as catering, security, fabrication, and houseboat should be done by indigenous companies. There is big clash between host communities and oil companies. Community challenge still exist in deepwater. |
Negative Comments	
Question C	What are your thoughts on treatment of host communities in the Niger Delta as it relates to oil and gas extraction?
Positive Comments	International oil companies go beyond their corporate responsibility in terms of investment in the communities. The oil companies are trying their best.

Neutral	Welfare disparity between host community's settlement and oil company's settlement. Government has not been present to provide security and good policy. Communal clashes and destruction of properties put in place by the oil companies.
Negative Comments	Impact of oil companies on the environment has been devastating. Government and oil companies have not been fair to the host communities. Some community people are lazy. Corruption issues where community leaders and representatives enrich themselves with largesse from the oil companies. Money given to help communities is not reaching the required area of needs. Criminality issue in the community. Pollution in the environment through bunkering, blowing up of pipelines and destructive activities of criminals in the communities.
Question D	How well do you think the nation can utilize deepwater gas deposit for developmental purposes?
Positive Comments	Deepwater can provide gas for electricity. Gas provide cleaner energy and thereby has prominent role in the future. It could be used as a springboard for growth. There is need to create a market. For instance, a policy that all government officials will use gas run vehicles will create demand for gas. Dangote trucks are now using gas. Some heavy-duty industrial generators are now being converted to utilize gas. Natural gas is used in fertilizer production.
Neutral	How are other places in the world able to utilize gas? There is need to put a policy in place and government should act. Policy

	enforcement is key. How can community enforce a particular policy? Who will be held responsible?
Negative Comments	Price cannot be controlled but we can control industrialization through extra policy initiative. Government's focus is more on generating money instead of creating infrastructure. Money from sales of oil and gas are not ploughed back into infrastructure development.
Question E	What other comment or remark do you have concerning the deepwater gas deposit in Niger Delta?
Positive Comments	Creating an enabling environment for deepwater gas is key, just like every segment of the economy. Deepwater gas reserve is huge. We should exploit it and make money in the long term. Government to seek alternative funding as a financing option with proper in-house re-orientation. The Nigerian Content Development Management Board (NCDMB) has seen the emergence of local companies that has financial backing and technical support from oversea companies. Gas to power project will go a long way.
Neutral	Holistic solution such as whistleblowing policy is necessary to keep people in check. Government should be willing to cooperate with the multinationals. Government officials must be willing to build on others legacy rather than scuttling the process of continuity. It is important to put the right people in right positions to guarantee performance.

	Market forces and prices are key issues to be addressed. The PIB did not specify about gas prohibition or flares out. Our various policies and laws need clarity around provisions that enabled foreign investors.
Negative Comments	Consistency in policy is a problem due to instability in government. This serves as a deterrent to prospective investors from bringing in their resources. Some of governments policies are too stringent to implement practically. Sabotage and nepotism are also dangers to proper development. There was a gas masterplan conceived many years ago which has not been effectively implemented because of changes with successive governments of different regimes.

Oil industry focus group. There were ten participants in the oil company focus group. Table 5 includes focus group responses from this group as a narrative summary and analysis.

Table 5

Discussion Responses by Oil Industry Focus Group

Question A	What is your opinion about government policy on gas deposits in the Niger Delta?
Positive Comments	Terms exist for onshore operations. Target on gas required for domestic consumption. Associated gas from deepwater operations are sent to Bonny NLNG (Nigerian Liquefied Natural Gas). Policy exists on individual onshore gas projects.
Neutral	Every resource belongs to government. Proceeds from Associated gas sales kept in escrow account.
Negative Comments	Nothing decisive on deepwater gas policy. No clear gas terms. Government has no solid position. Petroleum Industry Bill is undecided yet. No benefit for communities and country.
Question B	What do you think of the big oil companies' presence in deepwater Nigeria?
Positive Comments	A good thing for Nigeria. Partnering with indigenous companies helped with technology transfer. Value and experience were brought by multinationals. Deepwater is safe haven from community issues. Increased opportunity for employment.
Neutral	More multinationals are needed.
Negative Comments	Lots of acreages are underassessed. Room for more oil and gas projects. Funding challenge for indigenous operators. Over-reliance on international companies.

Question C	What are your thoughts on treatment of host communities in the Niger Delta as it relates to oil and gas extraction?
Positive Comments	Some benefits were provided: light, road, education, healthcare and safety.
Neutral	Social and economic impact. How will communities be impacted positively?
Negative Comments	Negative impact of oil extraction on communities. Irreversible environmental damage. No real benefit from oil and gas exploitation. Community leader's exploitation of followers. Shell exited from Warri. A wide gulf has been created. No visible result of social performance. Oil brought laziness. Government compensation to communities was unfair.
Question D	How well do you think the nation can utilize deepwater gas deposit for developmental purposes?
Positive Comments	Gas needed for power, industries and domestic use. Deepwater gas resource base is huge. Gas can be piped to onshore or Bonny Liquefied Natural Gas.
Neutral	Need to explore floating liquefied natural gas. The economy will blossom.
Negative Comments	There are no gas pipelines. Youth unemployment.
Question E	What other comment or remark do you have concerning the deepwater gas deposit in Niger Delta?
Positive Comments	Companies operation base is still onshore
Neutral	Need for legal clarifications.
Negative Comments	Legacy issues on community impact from deepwater activities.

All the oil industry respondents were affiliates of the oil and gas industry in Nigeria through either work, place of abode, or peculiar interest. All the respondents willingly shared their perspectives and opinions about deepwater gas deposits in the Niger Delta without any obvious bias. Two of the oil industry respondents were between the ages of 31 and 40; four each were between ages 41 and 50 and 51 and 60. There were nine male participants and one female participant. Nine respondents agreed that "the percentage of Nigeria's foreign exchange earnings from oil and gas was over 75%" and one respondent said it was between 51 and 75%. Six respondents agreed that gas reserves in deepwater Nigeria were greater than 300 trillion cubic feet; three respondents said they were between 200 and 300 trillion cubic feet, and one respondent said they were between 100 and 200 trillion cubic feet. Regarding the percentage of gas located in the deepwater, two respondents said it was less than a third; three said it was greater than two-thirds, and five respondents said it was between one third and two-thirds.

Six respondents said there were no government policies to exploit deepwater gas and four respondents had no opinion. Six respondents thought the government could not profitably monetize deepwater gas for export at the same pricing as onshore gas; three respondents thought it was possible, and one respondent was undecided. Nine respondents agreed the government could give adequate monetization policy incentives to help harness deepwater gas profitably; one respondent was undecided. Eight respondents noted a connection between deepwater gas and industrialization, but two respondents did not confirm this link. Nine respondents thought deepwater gas could help power steel, fertilizer, and petrochemical plants; one respondent was undecided. Eight respondents thought the Nigerian government would grant special incentives for deepwater gas monetization to support growth and industrialization; two respondents disagreed. All respondents thought the government could benefit from the global experience of multinational oil companies in determining deepwater gas pricing.

Seven respondents thought there would be real benefits to the communities from harnessing deepwater gas deposits; one respondent thought there was no benefit to communities and two respondents remained undecided. Nine respondents opined that the government would benefit most from harnessing deepwater gas deposits; one respondent thought multinational oil companies would benefit most. Nine respondents suggested that communities would benefit least from deepwater gas deposits; one respondent thought multinational oil companies would benefit least.

Analysis of oil company discussion. Given the predominant agreement on foreign exchange earnings, the consensus pointed toward higher exchange earning. This suggests an awareness of greater Nigerian earnings as understood by oil company personnel. Company personnel did not express a strong consensus on the level of gas reserves, which suggests that individuals were merely offering their personal estimates without official knowledge. This combination of certainty about earnings and uncertainty about the availability of reserves suggests that company personnel were not dependable regarding knowledge about the future value of gas reserves despite a strong sense of confidence about foreign exchange earnings.

The oil industry focus group opined that the federal government had no solid position on deepwater gas terms, which they interpreted as a possible rip-off of the country and host communities. Gas terms were clear for onshore operations, especially for individual gas projects. The federal government gave IOCs clear targets for quantities of gas required from various JV operations for domestic consumption. Deepwater gas exploitation, however, is different. Every mineral resource in the country statutorily belongs to the federal government. Participants reported that part of the gas policy was tied to the PIB, which was still under review by federal lawmakers without any date for its full implementation. Participants recommended that all gas extraction from deepwater should be channelled to Bonny LNG. The associated gas produced alongside operations in the deepwater could be kept in an escrow account until proper clarification was available. The implementation of associated gas policy and how much investment the government was willing to make into the resource was a subject of government indecision. Oil company personnel suggested that the government put measures in place for proper pricing of the deepwater gas to realize value from the resource.

The participants deemed the presence of big oil companies in Nigeria good development. More vested interest by multinational oil companies was desirable. Big oil companies performed well despite the challenges of deepwater oil and gas operations. There were significant numbers of oil prospecting leases (OPLs) and oil mining leases (OMLs) that were unassessed. There were opportunities for more oil and gas projects to mature in deepwater Nigeria based on participants' comments (e.g., "there is room for more oil and gas projects"). Many indigenous oil and gas companies swamped the landscape in the past decade. Participants recognized that some multinationals were successful in investing and realizing value from deepwater deposits; others were unsuccessful due to the significant amount of money they spent.

Deepwater ventures were inherently more expensive and only financially robust companies could undertake such exploration. Partnering with smaller indigenous companies boosted the Nigerian economy and brought better value propositions to all stakeholders. Multinationals brought a lot of value and experience to deepwater Nigeria projects. Participants noted that deepwater gas projects would grow once gas pricing issues are resolved. The deepwater provided a haven away from the community issues of onshore activities. Deepwater also provided employment opportunities, which benefit all stakeholders.

Oil company focus group participants noted social and economic effects of oil and gas extraction by the oil companies, which were mostly negative for communities. There were irreversible damages to the environment due to operations of oil and gas companies. Most of the areas of operation were remote and there was little effort to remediate damage to the environment. Participants opined that oil and gas exploitation did not benefit local communities. Leaders of the community did not allow the benefits to effectively filter down due to corruption. Participants stated that care for the community should be the government's responsibility, but oil companies need to manage the risks of their business to ensure sustainability. Oil companies should seek ways to transform communities so they might benefit from their activities. This could create real value and long-term positive impacts on the people and the environment.

Participants noted that oil companies created a disparity in living standards between company employees and the community. The oil and gas companies exploited the host communities because the communities did not realize the governmental requirements for social performance. Some benefits did reach communities (e.g., lights, roads, education, healthcare, and safety). Participants noted a negative impact on the psyche of the indigenous people of the Niger Delta. Some individuals in host communities accepted easy money from the oil and gas companies and misappropriated it. Corrupt leaders served as intermediaries, channelling benefits from oil and gas companies away from the host communities. Participants opined that the nation was lazy due to the boom of oil, typical of Dutch disease phenomenon (Chindo et al., 2014). There was a decline in agriculture and industry. The federal government determined ownership structures of mineral resources and compensation to communities was unfair.

Participants explained that Nigeria needed gas for power, industries, and domestic use; however, infrastructure (e.g., a pipeline from deepwater to facilitate this process) was lacking. Participants also noted that youth unemployment was a problem. They expressed that there was need to explore floating liquefied natural gas (FLNG) to yield foreign earnings. Companies could pipe the gas onshore for domestic use as well as to Bonny LNG.

Legal clarifications established deepwater as federal government territorial waters; they are not part of any Niger Delta community allocation. Participants described legacy issues regarding ways deepwater oil and gas activities impact communities, although they may not affect the day-to-day activities of the communities. The base of operations of oil companies was still onshore within the community even when all deepwater oil and gas extraction activities were offshore.

Government focus group. There were seven government focus group participants out of 11 potential participants. Six respondents confirmed their affiliation with the oil and gas industry in Nigeria either by work, place of abode, or peculiar interest; one respondent reported no affiliation. Six respondents said they were willing to share their perspectives about deepwater gas deposits in Niger Delta without any bias; one respondent did not make this claim. Three respondents were between the ages of 30 and 40; two respondents were in the age range 30 to 40, and two were between 50 and 60 years old. All respondents in the government stakeholder group were male. Three of the respondents had graduate degrees and four respondents had post-graduate degrees.

All respondents opined that over 75% of Nigeria's foreign exchange earnings came from oil and gas. Five respondents said Nigeria's gas reserves were between 100 and 200 trillion cubic feet; one respondent said they were between 200 and 300 trillion cubic feet, and one stated they were over 300 trillion cubic feet. All respondents said the proportion of gas deposit located in deepwater was between one third and two-thirds.

Five respondents said there was a government policy to exploit deepwater gas for profit; one respondent said there was no policy, and the other respondent did not know. Five respondents said the government could monetize deepwater gas for export at similar pricing as NLNG; two respondents disagreed. Five respondents said deepwater gas could be utilized in a similar manner to domestic utilization of onshore gas with different JV partners; one respondent disagreed, and the other remained undecided. All respondents believed the Nigerian government could give adequate incentives through monetization policy to foster the profitable harnessing of deepwater gas. All respondents noted a connection between gas utilization from deepwater and industrialization. Six respondents said the Nigerian government would be willing to grant special incentives for deepwater gas monetization to support growth and industrialization and one respondent disagreed. Five respondents said this project would help the government form a monetization policy; one respondent disagreed, and one remain undecided. All respondents believed that the government could benefit from the global experience of multinational oil companies in determining deepwater gas pricing.

Six respondents thought real benefits existed for communities from deepwater gas deposits; one respondent thought otherwise. Six respondents opined that the government would benefit most from harnessing deepwater gas, and one respondent said multinational oil companies would benefit most. All respondents agreed that communities would benefit least from harnessing the deepwater gas deposit.

Analysis of government discussion. Participants noted that the government policy on deepwater gas was still in progress and there was no effective policy in operation. Oil was the focus of exploration and production in the past and uncertainty existed about the real quantity of gas reserves. A working document on gas policy was still undergoing processing. They shared that the federal government was inconsistent regarding the basic tenets of locations of the oil and gas industry. Industries were in locations away from the Niger Delta due to the political interests of those in power rather than the requirements of the people. Participants noted that the government had under-harnessed gas in the Niger Delta to the extent that gas for power initiatives and fertilizer industries were politically motivated.

Gas terms should be in place as they are for JVs and PSCs for oil extraction. A working gas term should be a requisite condition for multinational oil companies to invest in the Niger Delta. Participants noted that natural gas policy was already approved by the Federal

Executive Council but not yet ready for implementation. The government gave the directive to maintain no flare from oil and gas operations in the Niger Delta but did not fully implement the policy directive. The government wanted to monetize gas for profit rather than flaring it, but there is no clear policy on this point.

Participants agreed that multinational oil companies were doing well in the Niger Delta but needed to increase ventures into the deepwater. This focus group considered the presence of big oil companies in deepwater Nigeria a rip-off because the PSC agreement of 1993 did not have clauses to ensure windfall from high oil prices would benefit the Nigerian government. Some of the multinationals were sitting on oil production facilities in the deepwater, making money without paying royalties.

Participants reported a lack of incentives for oil companies to search for gas because the PSC was based on oil production. Deepwater exploration is a very expensive venture and infrastructure development (e.g., building pipeline from deepwater to onshore) is equally expensive. Participants opined that gas industries wanting gas from deepwater at similar pricing to onshore gas would be unsuccessful without adequate incentives. The governance aspect of the PIB was a way to unbundle issues so that policies could be more specific.

Considering treatment of host communities, government participants felt the government had not fulfilled their expected level of service delivery. The government should have developed the host communities of the Niger Delta through provision of physical infrastructures. Cash payments made to leaders were a problem because corrupt community leaders misappropriated the money. Participants suggested training youths, building infrastructure, and improving communication with community members. Many community members lived in abject poverty while the neighboring oil company settlements enjoyed opulence. There was light, water, and good roads in the oil company settlements. Participants reported that provisions by the oil companies were exclusive of the surrounding community, which led to neglect of the environment and loss of livelihood for host communities. Before the mid-2000s, participants explained that host communities were mistreated but sustainable community development improved in later years.

Participants noted that most oil companies offered developmental projects; however, host communities were already prone to destructive behaviors (e.g., vandalism, stealing from pipelines). The government ensured oil producing communities engaged in decisions affecting their wellbeing, according to government personnel in this focus group. The government required that IOCs have projects to improve communities as part of their corporate social responsibilities. Participants stated that offshore benefits usually go to state governments, in accordance with a Supreme Court ruling.

Participants explained that if gas terms and policies were well constructed, the government could utilize deepwater gas deposits for developmental purposes. The nation needed to diversify efforts from oil and gas production to fertilizer and petrochemical industries, power generation, steel, vessel manufacturing, and green energy. Participants argued that enough gas existed onshore; deepwater gas was a secondary need that the nation was not adequately prepared to harness. They noted that creating an environment for multinationals to harness deepwater gas would help drive local industry. Nigeria was unable to harness enough gas to meet the WAGP project demands, and there was no infrastructural capacity to harness gas deposits onshore despite the need for power, export, and industrialization.

The government participants observed that there had been no attempt to estimate the gas reserves in the deepwater. After companies encountered non-associated gas while drilling for oil, they capped the reserves without further evaluation or determination of their usefulness. Participants noted upcoming projects to investigate the possible utilization of deepwater gas and suggested the government encourage the private sector to build infrastructure through incentives. However, such incentives for the IOCs in their exploration for oil allowed exploitation of policy loopholes in the selection of projects. For example, participants warned that the PSC was based on the collective knowledge of IOCs' consultants who supported policies that were highly beneficial to the oil companies in the long run. Table 6 includes focus group responses from government employees.

Table 6

Discussion Responses by Government Focus Group

Question A	What is your opinion about government policy on gas deposits in the Niger Delta?
Positive Comments	Government issued no flares directive. Driving towards monetization.
Neutral	Government policy is being worked upon. Oil was focus in the past.
Negative Comments	No clear policy. Industry location by government is inconsistent. Gas is under-harnessed. Gas utilization was politicized. Investors have no confidence.
Question B	What do you think of the big oil companies' presence in deepwater Nigeria?
Positive Comments	Big oil companies are doing well. They are in business for profit.
Neutral	Gas belonged to government.
Negative Comments	The 1993 Production Sharing Contract (PSC) was an oil-based rip-off. No incentives to explore for gas. Deepwater exploration is very expensive. No fiscal terms for gas.
Question C	What are your thoughts on treatment of host communities in the Niger Delta as it relates to oil and gas extraction?
Positive Comments	Creation of confidence through youth empowerment. Government required companies to improve their areas of operation.

Neutral	Offshore benefits statutorily belonged to state government.
Negative Comments	Government has not done well. No physical infrastructure development. Cash payment to community leaders were diverted. Disparity in living standards. Oil company provisions sometimes excluded community members. Unfair treatment of host communities.
Question D	How well do you think the nation can utilize deepwater gas deposit for developmental purposes?
Positive Comments	Diversification is essential. Fertilizer, petrochemical industry, power generation would be developed. Oil majors were business oriented. Gas could be utilized for power supply, export and industrialization.
Neutral	Gas terms and policies to be made right.
Negative Comments	Impending problem of clean energy demand. There was sufficient gas onshore, deepwater gas was secondary source. Lack of good fiscal terms was a deterrent. Vandalism of facilities.
Question E	What other comment or remark do you have concerning the deepwater gas deposit in Niger Delta?
Positive Comments	Upcoming projects to utilize some of the deepwater gas. Encouraged to build infrastructure.
Neutral	Government should invest massively in gas.
Negative Comments	No deliberate attempt to estimate deepwater gas reserves. They were not primary focus of exploration. Exploitation of policy loopholes by multinationals.

Results from all focus groups. A combination of responses from all participants appears as a narrative summary in the following section. The collation of results from all three focus groups of stakeholders made it easy to identify areas of consensus. Responses by a majority (80% or more) of the respondents were obvious and guided recommendations for action. All but one respondent had affiliations with the oil and gas industry in Nigeria either by work, place of abode, or peculiar interest and were willing to share their perspectives about deepwater gas deposits in the Niger Delta. Eleven respondents were in their 30s, ten respondents were in their 40s, and six respondents were in their 50s. There were 25 male respondents and 2 female respondents. Fifteen respondents had graduate degrees and 12 respondents had postgraduate degrees.

Twenty-four respondents opined that over 75% of Nigeria's foreign exchange earnings came from oil and gas; three respondents said it was between 51% and 75%. Ten respondents agreed that Nigeria's gas reserves were over 300 trillion cubic feet, nine respondents thought they were between 200 and 300 trillion cubic feet, and eight respondents stated they were between 100 and 200 trillion cubic feet. Twenty respondents said the proportion of gas deposit located in deepwater was between one third and two-thirds; five respondents said it was over two-thirds, and two respondents said it was less than a third.

Ten respondents affirmed that a government policy to exploit deepwater gas for profit existed; nine respondents said there was no policy and eight respondents were undecided. Seventeen respondents thought the government could profitably monetize deepwater gas for export at similar pricing as onshore gas; eight respondents did not believe it was achievable, and two respondents remained undecided. Twenty-one respondents thought deepwater gas could be utilized in a similar manner as onshore gas for domestic use; four respondents disagreed, and two respondents were undecided. Twenty-six respondents thought the Nigerian government could give adequate incentives through monetization policy to help harness deepwater gas, and one respondent remained undecided. Twenty-five respondents agreed there was a connection between gas utilization from deepwater and industrialization; two respondents did not. Twenty-five respondents thought deepwater gas could help power steel, fertilizer, and petrochemical plants; one respondent disagreed and another was undecided. Twenty-two respondents thought the government would be willing to grant special incentives for deepwater gas monetization to support growth and industrialization; one respondent thought otherwise and four remained undecided. Twenty-three respondents thought this research would help the government create a policy; one respondent thought it would not and three remained undecided. All respondents thought the government could benefit from the experiences of multinational oil companies in determining deepwater gas prices.

Twenty-two respondents thought there would be real benefits to host communities from harnessing deepwater gas deposits; three respondents thought otherwise and two remained undecided. Twenty-three respondents opined that the government would benefit the most from harnessing deepwater gas deposits; four respondents suggested that multinationals would benefit most. Twenty-five respondents stated that communities would benefit least from harnessing deepwater gas deposits; two respondents argued multinationals would benefit least. Table 7 includes details of all responses from research participants for the survey and focus group combined.

Table 7

Combined Survey Responses of Participants

Questions	Answers		
1a. Do you have any affiliation to the oil and gas industry in Nigeria either by work, place of abode or peculiar interest?	Yes 26	No 1	
1b. Are you willing to share your perspectives about deepwater gas deposits in Niger Delta without any bias?	Yes 26	No 1	
1c. Age	31-40 11	41-50 10	51-60 6
1d. Gender	Male 25	Female 2	
1e. Education	High School 0	Graduate 15	Post graduate 12
2a. About what percentage of Nigeria's foreign exchange earning comes from oil and gas?	26-50% 0	51-75% 3	>75% 24
2b. About how much gas reserves does Nigeria have?	100-200tcf 8	200-300tcf 9	>300tcf 10

2c. About what percentage of the gas is located in deepwater Niger Delta?	<1/3	1/3 - 2/3	>2/3
	2	20	5

3. Is there a government policy to exploit the deepwater gas for profit?	Yes	No	Don't know
	10	9	8

4. Government has the Nigerian Liquefied Natural Gas (NLNG) trains that have been exporting gas for profit. Do you think government can profitably monetize the deepwater gas for export at same pricing?	Yes	No	Don't know
	17	8	2

5. Government (in affiliation with her partners in the joint venture) has been utilizing gas from onshore for domestic uses. Do you think the deepwater gas can be utilized in a similar manner?	Yes	No	Don't know
	21	4	2

6. Gas from deepwater is typically more expensive to harness than onshore. Do you think the Nigerian government can give adequate incentives through monetization policy to help harness the gas profitably?	Yes	No	Don't know
	26	0	1

7. Do you see any connection between gas utilization from deepwater and industrialization?	Yes	No	Don't know
	25	2	0

8. Do you think the deepwater gas can help power the steel industry, the fertilizer and petrochemical plants?	Yes	No	Don't know
	25	1	1

9. Do you think Nigerian government will be willing to grant special incentives for deepwater gas monetization to support growth and industrialization?	Yes	No	Don't know
	22	1	4

10. Do you think this kind of research with opinions from oil industry, government and community personalities could help government decide her monetization policies?	Yes	No	Don't know
	23	1	3

11. Do you think government can benefit from global experience of multinational oil companies in determining deepwater gas pricing?	Yes	No	Don't know
	27	0	0

12. Do you think there is real benefit to the communities from harnessing the deepwater gas deposits?	Yes	No	Don't know
	22	3	2

13. In your opinion, which of the three groups below will have the most benefit from harnessing the deepwater gas deposits?	Multi-national oil companies	Govt. (fed./state/ local)	Communities in the Delta
	4	23	0

14. In your opinion, which of the three groups below will have the least benefit from harnessing the deepwater gas deposits?	Multi-national oil companies	Government	Communities in the Delta
	2	0	25

Findings

Comparison of the survey results and discussions from the three different focus groups via narrative analysis resulted in the identification of common areas of interest (possible consensus) and areas of disagreement that require further articulation. By itemizing areas of common interest in the focus group discussions that related to common themes, the creation of specific recommendations was possible. The assessment of results from the questionnaire surveys and focus group oral discussions facilitated a thematic identification of common issues. The following sections include the findings for each theme.

Establishing the stakeholder. All but one participant had affiliations with the oil and gas industry in Nigeria and freely gave their opinion about deepwater gas deposits in the Niger Delta. The single exception was a participant from the government focus group. There was an equitable range in age from 30 to 50 years old, which portends a balance in both outlook and experience of the participants. There were predominantly male respondents, which reflects the nature of this highly technical industry in which women are in a minority. Educational levels of graduate and post-graduate degrees confirmed the cognition level of respondents.

Knowledge about gas deposits and government policy in deepwater. Most respondents confirmed the dependence of Nigeria on foreign earnings from oil and gas. The quantity of gas reserves varied in almost equal amounts between the three focus groups. Most government focus group members thought the reserves were 100 to 200 trillion cubic feet, most community focus group members thought they were 200 to 300 trillion cubic feet, and most oil industry focus group members thought they were over 300 trillion cubic feet. The perceptive differences in the gas reserves of the different focus groups could be a source of dissention among these groups. Reserves in the ground typically translate to resources and wealth for different stakeholders that benefit from oil and gas extraction.

Participants' opinions about the existence of government policy on deepwater gas also varied. Both the oil industry and community focus groups agreed there was no policy, but members of the government focus group argued there was a policy. Further comments by both community and oil industry focus group members suggested that government policy was indecisive and inconsistent; the government focus group stated that the policy was in progress and monetization of gas through this policy was unclear because the policy was unfinished. One third of the respondents did not know about any gas policy for deepwater. Members of community and government focus groups thought that the government could monetize deepwater gas for export in the same way as the LNG from onshore, but members of the oil industry focus group disagreed. This difference in opinion clarified the dichotomy of perspectives about policy regarding monetization of deepwater gas in the Niger Delta.

Deepwater gas utilization. Members of all three stakeholder groups agreed that the government could use the deepwater gas deposit for domestic purposes in a similar way as onshore gas. This followed the understanding that policy for onshore gas utilization existed in various JV agreements between the government and oil companies. There was also a consensus that the government could give adequate incentives through monetization to harness deepwater gas. Inherent in this response was the consideration that deepwater gas was more expensive to harness than onshore gas. The JV partners harnessing onshore gas could determine the viability of each project before embarking on them via established government policy for onshore gas. The demand for gas was known, pricing was predictable, and resource availability was determinable.

Most respondents from all three stakeholder groups recognized a connection between deepwater gas and industrialization. Most thought deepwater gas could help power steel, fertilizer, and petrochemical industries. The community focus group emphasized issues plaguing the utilization of deepwater gas such as the need for a workable policy and infrastructure development. The oil industry focus group noted youth unemployment as part of the problem with deepwater gas utilization. The government focus group noted counterproductivity, citing infrastructure for gas development away from the Niger Delta for political reasons. The effectiveness of infrastructural investment would not be easy to realise unless these groups reach a consensus at the onset of policy-making.

Monetization policy. Most respondents from all stakeholder groups thought the government would be willing to grant special incentives for deepwater gas monetization to support growth and industrialization. The understanding that deepwater gas belonged exclusively to the government made stakeholders wary; they had to seek special incentives to participate in extracting the resources for value. Members of the community focus group believed that re-injecting and utilizing flared gas was a way to create profit. Deepwater extraction provides some safety from community issues;

therefore, the path to monetization may not be fraught with the same challenges. Most respondents thought this project could help the government form better monetization policies for deepwater gas reserves. The content of special incentives varied due to the variety of activities within the value chain. The desire to reach consensus to craft an all-encompassing policy for monetizing deepwater gas could help unlock the economic potential of deepwater gas reserves.

Benefits of deepwater gas. All respondents thought that the government could benefit from the global experiences of multinational oil companies in determining deepwater gas pricing. Gas was an internationally traded resource, and international assessment of value of the resources in deepwater Nigeria was an important element of monetization policy. Most respondents thought that harnessing deepwater gas deposits would benefit local communities. Nonetheless, most argued that the government (federal/state/local) would benefit most from harnessing the deepwater gas deposits; communities would benefit least.

Summary

Most respondents confirmed the dependence of Nigeria on oil and gas reserves, but opinions varied regarding the quantity of deepwater gas reserves that are present. There was no agreement about existing government policy concerning deepwater gas reserves and no clear path to monetization. This suggests an opportunity for further consultations and dialogue. There was consensus that provision of government incentives would enable proper exploitation of deepwater gas for domestic use despite the higher costs of extraction. Most respondents noted a connection between deepwater gas extraction and industrialization in the power, steel, fertilizer, and petrochemical industries. Infrastructural development for profitable evacuation of deepwater gas was an essential element of a workable monetization policy. Deepwater gas belongs to the government and its extraction may be void of the community disturbances that onshore gas projects experience. A consensus on policy to monetize deepwater gas would be beneficial given the immense value of gas to all the stakeholders in the gas value chain.

CHAPTER 8

FINAL THOUGHTS

The deepwater gas industry in Nigeria currently has no policy in place for profitable extraction of this natural resource. There are various issues associated with extraction of oil and gas in Nigeria that require evaluation and appropriate management before stakeholders can arrive at a working policy for extraction. Key stakeholders (i.e., community, government and oil industry members) have the knowledge to drive the possible implementation of a working policy to help monetize deepwater gas deposits. Survey and focus group discussions with these stakeholders revealed common themes of consensus opinions on policy proposal regarding deepwater gas deposits in the Niger Delta. Consensus among key stakeholders may facilitate the creation of a policy framework for monetization of deepwater gas reserves in the Niger Delta that benefits all groups. Sharing these consensus opinions with policy makers may positively influence the creation of a working policy for deepwater gas extraction.

Most participants in the study perceived the government to benefit most from oil and gas resources while communities benefited least. Special incentives are necessary to encourage multinationals to participate in deepwater gas projects in partnership with indigenous companies. The government policy for sharing oil wealth was unsatisfactory, according to most respondents. Policy for gas extraction from deepwater must correct inequalities and improve environmental protections. Policy incentives for deepwater gas exploitation may enable more accurate determination of gas reserves in deepwater Nigeria. Provision of adequate security and favorable government policy could curb criminality, vandalism, communal clashes, and corruption in the Niger Delta.

Results

Stakeholders meaningfully contributed to this case study. They demonstrated knowledge about deepwater gas deposits and government policy in the Niger Delta. An integration of questionnaire survey and focus group results revealed areas of agreement, common interests, and disagreements. Infrastructural development for gas extraction is lacking in the deepwater regions of the Niger Delta. Deepwater gas could benefit all stakeholders, which was an important driver for the creation of an attractive monetization policy. Navigating policy issue differences among the interested stakeholders to arrive at an agreeable policy through consensus is a worthwhile pursuit in the peculiar Nigerian terrain. A policy for deepwater gas monetization could provide great financial potential for all stakeholders in the deepwater gas venture.

Deepwater gas could provide domestic benefits in a similar fashion to onshore gas. Utilization of deepwater gas for the power of fertilizer, steel, and petrochemical industries were all feasible and valuable proposals. There is a need for infrastructure development targeted at deepwater gas. This investment could create job opportunities for youths who may then resist vandalization of oil and gas installations. Community activism, vandalism, and disruption of operations would decrease if the extraction of deepwater gas resulted in profitable communal settlements.

Participants agreed that there was currently no policy for monetizing deepwater gas deposits in the

Niger Delta; however, the government was actively working on a practicable policy. Deepwater gas is more expensive to harness than onshore gas because there is no infrastructure for the extraction of gas. Government policy, in deepwater specifically, addressed the extraction of oil and excluded gas. Most participants agreed that special incentives would be necessary in government policies to encourage multinational oil companies to engage in deepwater gas exploitation. A few indigenous corporations have equity shares, but most shares and operatorship remain with multinationals. Continued operation in deepwater should include gas exploitation, which could improve deepwater technology transfer and active participation by indigenous companies.

Members of the community stakeholder group were well-versed in issues pertaining to the community where they witnessed the influence of oil and gas operations. Their responses demonstrated adequate knowledge of the devastating effects of oil and gas operations in their communities with a keen interest in proffering a lasting solution. The oil industry stakeholder group expressed a desire to maintain regulations in extraction of oil and gas to ensure minimal environmental and human impact. The quest for sustainability showcased the tenets of stakeholder theory; satisfying other key stakeholders protected their interests as business people. Various community assistance programs reflected the oil industry stakeholder group's commitment to sustainable development in areas of operations. The level of corruption experienced in their efforts to help communities was a frustrating roadblock that elicited tripartite agreement of key stakeholders to ameliorate the problem. Government personnel explained that policy was available for gas extraction onshore and they were currently working on deepwater gas policy. The delay in formulating a policy was a demonstration of their interest in maximizing profit from the deepwater gas. Participants suggested that the government may want to use policy to accumulate more benefits for the nation from deepwater reserves.

The respondents from the government focus group were knowledgeable about government policy that was in-progress. They expressed personal opinions about government control of subsidiary organizations to benefit the general populace. This is one reason the government did not grant a gas monetization policy at the beginning of the exploration venture in Nigerian deepwater in the mid-1990s. The interest of the government was to harness maximum benefits for the public through a favorable monetization policy. The agency's monitoring of compliance was effective and staff were aware of the selective execution of profitable projects by multinationals in the onshore acreages where gas extraction policy existed. Some respondents confirmed the government's hesitation to enact policy for exploitation of deepwater gas without maximizing benefits for the public.

Interpretation of the Findings

The potential for deepwater gas to positively affect the economy in the same manner as onshore gas has been overlooked over the years. There has been extraction of associated gas produced from deepwater during the process of oil extraction. Contractual agreements in the PSC stipulate a channelling of such extracted gas for monetary benefit, which proved that deepwater gas could be extracted for profit. Nigeria is in dire need of stable power generation and deepwater gas could be a ready resource in a solution-oriented gas-to-power initiative. This study demonstrates the advantages of research that brings workable solutions to long-standing problems in the country. Some gas-powered industries already take supplies from onshore gas plants. This implies great potential for increases in industrialization with the possible harnessing of huge gas deposits in deepwater for fertilizer, steel, and petrochemical industries. The guarantee of a steady, environmentally-friendly gas source from deepwater is another incentive for advantageous extraction for industrialization. To realize this industrialization, targeted infrastructure development of deepwater gas will be a necessary investment. Such investment may boost job creation, technology transfer, and local capacity building. The spate of disruptions to oil and gas operations by the communities are exacerbated by widespread corruption in many of the settlement offers made to the communities by the oil industries. Recommended infrastructural development may create jobs and ease tensions with youths to curb the vandalization of oil and gas installations. This may lead to attendant stability in operations that further guarantee stable income for the benefit of stakeholders and the nation at large.

There is currently no policy for monetization of deepwater gas even though the government is working on a realistic policy. This presents an opportunity for an inclusive consultation of key stakeholders' interests in a workable policy. Deepwater gas is more expensive to harness than onshore gas; the need to create infrastructure will depend on foreign investments if necessary incentives are available. This will also create job opportunities and technology transfer that will boost the economy. Providing a solution to the government regarding the opportunity for deepwater gas extraction through workable policy initiatives will create several special interests both in the government and private sectors. This process may include the collaboration of indigenous companies in bidding for deepwater projects that were traditionally the domain of multinationals, the benchmarking of resource-to-value in comparison with other oil producing nations, and diversification from an oil-based economy though development of ancillary service sectors of the economy. The setting of safety standards by multinationals in oil and gas operations has multiplier effects in other areas of the economy including healthcare and environmental well-being.

There are a few limitations to this study that provide opportunities for possible improvement and areas for further research in the future. First, a change in government could redefine operating conditions and terms governing gas commercialization policies in the deepwater. The impact could be lack of interest in monetizing the deepwater gas. Therefore, future research should address the specific regime in place at the time. Second, the passing of the PIB for modernizing procedures in the oil and gas industry in general could increase interest in deepwater gas pricing regulation and make it difficult to agree on common grounds for monetizing the deepwater gas. Another concern for future researchers is that the level of cooperation from participants may change; community and oil industry participants were forthcoming but government employee participants were not as available. Additionally, the downturn in the industry due to a crash in oil prices and consequent recession in Nigeria raised constraints on organizations, communities, and individuals. Many agriculture-based initiatives were government-supported alternatives to the oil and gas economy. Consideration of the recommendations from this study by government policy makers is uncertain. The DPR will receive a copy of the report, but there is no guarantee that these perceptions of deepwater gas monetization will reach policy-makers or influence policy decisions.

Recommendations

The government has a responsibility to generate workable policies for profitable extraction of deepwater gas reserves. The government must consult stakeholders regarding their interests. The system of governance in Nigeria, with grassroots representations at the seat of government, should include candid representation of community interests; however, observations and feedback from focus group discussions indicated a huge gap in this area. Dialogic engagement with the government may improve the development of a working monetization policy for deepwater gas reserves in the Niger Delta. The government should embark on infrastructural development of gas extracting facilities in the deepwater Niger Delta. The lack of gas-oriented infrastructure is a disincentive for multinationals. The discussion of the government focus group suggested that the government has a gas masterplan for onshore gas development; it should include deepwater gas to align the demand for infrastructural development with a holistic pursuit of deepwater gas extraction.

The oil industry focus group revealed a trend in community development; individuals from communities cause havoc for oil and gas installations through vandalism and illegal oil extraction activities due to ineffective community assistance and development programs. The government should foster long-term reversal of this trend via a water-tight tripartite agreement between the government, oil industry, and community councils that engages responsible community representatives to help solve the problem. The psyche of the community of the Niger Delta suffered due to the degradation effects of oil and gas exploitation activities. A viable path to psycho-social recovery must include government support to remediate environmental damages.

Conclusions

The significance of this research is the insight it offers to policy makers about consensus among seemingly disparate stakeholders regarding policy proposals for the monetization of deepwater gas reserves in the Niger Delta. The government (that may use new perspectives to influence policy), the oil companies (that may benefit from a new policy), and the community (that may benefit more from new resource extraction than they have in the past) may benefit from this study. The articulation of issues during focus group discussions helped elucidate some of the most debilitating problems of oil and gas exploitation in the

Niger Delta. Collation of questionnaire survey results and transcribed focus group discussions resulted in thematic assessment of issues concerning deepwater gas monetization.

The three key stakeholder groups ensured a robust appraisal of the issues surrounding deepwater gas deposits in the Niger Delta. Data from the questionnaire survey and focus group discussions provided comprehensive qualitative responses that reflected the opinions of key stakeholders. Companies in Nigeria extracted gas from onshore locations for both domestic use and export through NLNG trains. A significant portion of the total discovered gas resources is in deepwater Niger Delta. The PSC agreement between the Nigerian government and multinational oil companies operating in the deepwater exclusively focused on oil. There is no policy to extract and monetize deepwater gas. Difficulties in establishing deepwater gas projects include insurgency, media and international activism, bunkering, piracy, kidnappings, environmental degradation, oil spills, gas flaring, livelihood eradication, and institutionalized corruption. These factors drastically affect public policy for commercial exploitation of the sizeable deepwater gas resource in the Niger Delta. A consensus approach to policy to monetize deepwater gas would be an ideal outcome of this study.

Based on the consensus of all respondents, there was currently no policy for monetizing the deepwater gas deposits in the Niger Delta. The government stakeholder group suggested the government was working on a practicable policy. Some respondents in the government focus group noted a preferential execution of the available onshore gas extraction policies by multinational oil companies based on lopsided benefits for business people to the detriment of the Nigerian populace. Some respondents opined that the international influence of multinationals in crafting the PSC tilted benefits in their favor. The opinion of most participants was that the government would benefit most from harnessing deepwater gas and communities would benefit least. This mirrors the sharing formula of oil wealth by the Nigerian government, much to the dissatisfaction of different sectors of the economy.

Members of various focus groups encouraged participation of multinationals in deepwater gas extraction to enable technology transfer and eventual involvement of indigenous companies. They noted that onshore oil extraction increased indigenous company participation after acreage divestment exercises of marginal fields. The investment created jobs for the community and improved the complex process of oil and gas exploitation.

Most respondents confirmed that over 75% of Nigeria's foreign exchange earnings came from oil and gas. Gas exploitation was only a tiny fraction of expected capacity utilization for a populous, developing country like Nigeria. Most respondents agreed there was a connection between gas utilization from deepwater and industrialization. The community focus group emphasized the need for the government to build infrastructure for harnessing deepwater gas. Effective gas sector development could be catalyst for growth with multiplier effects on many sectors of the economy. Most participants thought that deepwater gas could help power the steel, fertilizer, and petrochemical industries. The government has exclusive rights to deepwater gas, and the enactment of workable policy based on a consensus between stakeholders would improve deepwater gas development.

Members of various discussion groups said that there was no development of a deepwater gas infrastructure to promote extraction and utilization. The government should provide special incentives to encourage commitment by multinationals to facilitate development of gas infrastructure in the deepwater Niger Delta. Environmental and health hazards of gas flaring experienced in onshore oil and gas extraction would occur less in deepwater gas exploitation due to absence of immediate communities next to extraction facilities.

Many respondents in the focus groups attributed the technical capabilities of indigenous oil and gas companies to the process of technology transfer from multinational companies. Discussions from the focus groups about partnerships between multinationals and indigenous companies in various JV arrangement epitomized the benefits of integrating infrastructural development with attractive policy incentives for technology transfer. Participants noted that many communities still lived in abject poverty. The community focus group articulated the impact of oil spills, gas flares, effluent discharge, contamination, biodiversity, and loss of livelihood due to the operations of oil and gas companies. A monetization policy should address the negative impact of oil and gas operations on community members and the environment. The rise of criminality in the Niger Delta may decrease with provision of an appropriate monetization policy.

Members of the community focus group identified corruption as the main issue relating to poor treatment of host communities in the extraction of oil and gas. The multinationals invested in welfare of the host communities, but community representatives and leaders were more interested in personal gain. The escalation of disagreements led to communal clashes and destruction of some properties built in the communities by the oil companies. Participants also observed shortcoming in the government's ability to provide adequate security and favorable policy.

The government stakeholder group exhibited hierarchical authority in governance when preparing for the discussion (Lynn & Robichau, 2013). There was a deference by some of the respondents to their superiors regarding where to conduct the meeting for convenience. A junior officer should not openly oppose his superior in government offices. A government staff member politely declined the invitation to participate because of her present involvement on some aspect of government work that her superiors may find incompatible with her opinions.

The sharing formula for oil wealth by the Nigerian government was unsatisfactory for many sectors of the economy. Members of the government group corroborated the view that the government would benefit most from harnessing deepwater gas and communities would benefit least. Participants stated that policy makers would typically protect their own interests when adjudicating rights and privileges; this resulted in marginalization of communities in oil wealth distribution by successive governments (Hilmer, 2010).

There was a marked absence of indigenous businesses in the oil and gas extractive activities in the Niger Delta. Members of various focus groups encouraged participation of multinationals in deepwater gas extraction to enable technology transfer and eventual participation of indigenous companies. Onshore oil extraction increased indigenous company participation in recent years. Government policy facilitated deepwater oil extraction through PSC rather than previously operated JOAs for onshore oil and gas extraction (Ogbonna & Ebimobowei, 2012). Some respondents from the government focus group observed a preferential execution of these policies by multinational oil companies. Some respondents opined the international influence of multinationals in crafting the PSC.

Effective gas sector development may be a catalyst for growth in the economy with multiplier effects on many sectors of the economy. Most of the respondents recognized a connection between gas utilization from deepwater areas and industrialization and thought that deepwater gas could help power industries. Participants argued that the government must provide special incentives to encourage multinationals to invest in gas infrastructure in deepwater Nigeria. An incentivised policy for extraction of gas would also enable a more accurate determination of gas reserves in deepwater Nigeria.

Development of indigenous entrepreneurial capabilities in extractive technologies requires partnering with multinationals (Nwoke, 2016). Many of the respondents in the focus groups attributed the technical know-how of indigenous oil and gas companies to technology transfer from multinationals. Closing the energy utilization gap through partnership is an effective strategy (Lester & Hart, 2015). The capital-intensive nature of deepwater gas exploitation further confirmed the need for collaborative partnership for sustainable monetization. A consensus of key stakeholders is essential to create a favorable policy framework for monetization of deepwater gas reserves in the Niger Delta.

Gas flaring leads to health problems in communities and poor agricultural yields. An equitable monetization policy should address gas flaring to positively affect the livelihood of community members and the environment. Oil spills contaminated the environment due to pipeline vandalization and aging facilities (Ordinioha & Brisibe, 2013). The community focus group emphasized the need for the government to build an infrastructure for harnessing deepwater gas to ensure oil and gas resources are well-contained and no longer harm the environment.

Corruption undermined oil companies' efforts to invest in the welfare of host communities. The community focus group noted that multinationals go beyond their corporate responsibility to invest in communities. Imbalance of power relations fuels upheavals in the community (Odoemene, 2012). The community focus group noted that the government and oil companies were unfair when addressing host communities' issues. Community members must have opportunities to express their concerns to improve policy on monetization of deepwater gas reserves.

Special incentives are necessary to encourage multinationals to participate in deepwater gas projects in partnership with indigenous companies. The government policy for sharing oil wealth was unsatisfactory, according to most respondents. Policy for gas extraction from deepwater must correct inequalities and improve environmental protections. Policy incentives for deepwater gas exploitation may enable more accurate determination of gas reserves in deepwater Nigeria. Provision of adequate security and favorable government policy could curb criminality, vandalism, communal clashes, and corruption in the Niger Delta. Consensus among key stakeholders may facilitate the creation of a policy framework for monetization of deepwater gas reserves in the Niger Delta that benefits all groups.

References

Abiola, J. O., & Ashamu, S. O. (2012). Environmental management accounting practice in Nigeria National Petroleum Corporation (NNPC). *European Scientific Journal, 8*(9), 76.

Agbiboa, D. E., & Maiangwa, B. (2012). Corruption in the underdevelopment of the Niger Delta in Nigeria. *Journal of Pan African Studies, 5*(8), 108-132.

Aigboduwa, J. E., & Osaimoje, M. D. (2012). Promoting small and medium *enterprises in the Nigerian oil and gas industry. European Scientific Journal, 9*(1), 244.

Ajugwo, A. O. (2013). Negative effects of gas flaring: The Nigerian experience. *Journal of Environmental Pollution and Human Health, 1*(1), 6-8.

Aminu, S. A. (2013). The militancy in the oil rich Niger Delta: Failure of the federal government of Nigeria. *Interdisciplinary Journal of Contemporary Research in Business, 4*(11), 813.

Atsegbua, L. A. (2012). The Nigerian oil and gas industry content development act 2010: An examination of its regulatory framework. *OPEC Energy Review, 36*(4), 479-494.

Baxter, P., & Jack, S. (2008). Qualitative case study methodology: Study design and implementation for novice researchers. *The Qualitative Report, 13*(4), 544-559.

Blackman, T., Wistow, J., & Byrne, D. (2013). Using qualitative comparative analysis to understand complex policy problems. *Evaluation, 19*(2), 126-140.

Brammer, S., & Jackson, G. (2012). Corporate social responsibility and institutional theory: New perspectives on private governance. *Socio-economic Review, 10*(1), 3-28.

Bridoux, F., & Stoelhorst, J. W. (2014). Micro-foundations for stakeholder theory: Managing stakeholders with heterogeneous motives. *Strategic Management Journal, 35*(1), 107-125.

Campbell, J. L., Quincy, C., Osserman, J., & Pedersen O. K., (2013). *Coding in-depth semi-structured interviews: Problems of unitization and intercoder reliability agreement.* Thousand Oaks, CA: Sage Publications.

Cao, G., Clarke, S., & Lehney, B. (2003). Diversity management in organizational change: Towards a systemic framework. *Systems Research and Behavioral Science, 20*, 231-242.

Chan, M. C., Watson, J., & Woodliff, D. (2014). Corporate governance quality and CSR disclosures. *Journal of Business Ethics, 125*(1), 59-73.

Chindo, M., Naibbi, A. I., & Abdullahi, A. (2014). The Nigerian extractive economy and development. *Human Geographies, 8*(2), 71.

Curry, N. (2013). Planning and policy documents as transactions costs: The case of rural decision-making in England. *Land Use Policy, 30*(1), 711-718.

Demir, T., & Reddick, C. G. (2012). Understanding shared roles in policy and administration: An empirical study of council-manager relations. *Public Administration Review, 72*(4), 526-536.

Department of Petroleum Resources (DPR). (2016). Official website of DPR. Retrieved from http://www.dpr.gov.ng

Dias-Sardinha, I., & Reijnders, L. (2001). Environmental performance evaluation and sustainability performance evaluation of organizations: An evolutionary framework. *Eco-Management and Auditing, 8*, 71-79.

Doos, M., & Wilhelmson, L. (2014). Proximity and distance: Phases of intersubjective qualitative data analysis in a research team. *Quality & Quantity, 48*(2), 1089-1106.

Edino, M. O., Nsofor, G. N., & Bombom, L. S. (2010). Perceptions and attitudes towards gas flaring in the Niger Delta, Nigeria. *The Environmentalist, 30*(1), 67-75.

Ekhator, E. O. (2015). Regulating the activities of oil multinationals in Nigeria: A case for self-regulation? *Journal of African Law, 60*(1), 1-28.

Eweje, G. (2006). Environmental costs and responsibilities resulting from oil exploitation in developing countries: The case of the Niger Delta of Nigeria. *Journal of Business Ethics, 69*(1), 27-56.

Ezeani, E. C. (2012). Economic and development policy-making in Nigeria. *Journal of African Law, 56*(1), 109-138.

Garrouste, P., & Brousseau, E. (2011). Institutional changes: Alternative theories and consequences for institutional design. *Journal of Economic Behavior and Organization, 79*(1/2), 3-19.

Goderis, B., & Malone, S. W. (2011). Natural resource booms and inequality: Theory and evidence. Scandinavian *Journal of Economics, 113*(2), 388-417.

Groves, H. (2005). Offshore oil and gas resources: Economics, politics and the rule of law in the Nigeria-Sao Tome E Principe joint development zone. *Journal of International Affairs, 59*(1), 81-156.

Gujba, H., Mulugetta. Y., & Azapagic, A. (2010). Environmental and economic appraisal of power generation capacity expansion plan in Nigeria. *Energy Policy, 38*(10), 5636-52.

Gullick, J., & West, S. (2012). Uncovering the common ground in qualitative inquiry. International *Journal of Health Care Quality Assurance, 25*(6), 532-548.

Hilmer, J. D. (2010). The state of participatory democratic theory. *New Political Science, 32*(1), 43.

Hsu, C., & Chang, P. (2013). Innovative evaluation model of emerging energy technology commercialization. *Innovation, 15*(4), 476-483.

Ibaba, S. I. (2008). The SPDC and sustainable development in the Niger Delta. *International Journal of Development Issues, 7*(1), 41-55.

Idemudia, U., Cragg, W., & Best, B. (2010). The challenges and opportunities of implementing the integrity pact as a strategy for combating corruption in Nigeria's oil rich Niger Delta region. *Public Administration and Development, 30*(4), 277-290.

Idemudia, U., & Ite, U. E. (2006). Corporate-community relations in Nigeria's oil industry: Challenges and imperatives. *Corporate Social Responsibility and Environmental Management, 13*(4), 194-206.

Ilegbinosa, I. A. (2013). An appraisal of fiscal policy measures and its implication for growth of the Nigerian economy: 1970-2009. *Advances in Management and Applied Economics, 3*(4), 193.

Jensen, T., & Sandstrom, J. (2011). Stakeholder theory and globalization: The challenges of power and responsibility. *Organizational Studies, 32*(4), 473.

Kefela, G. T. (2012). Organizational culture in leadership and management. *Project Management World Today, 14*(1), 1-12.

Kizito, E. U. (2014). The nexus between tax structure and economic growth in Nigeria: A prognosis. *Journal of Economic and Social Studies, 4*(1), 107.

Knudsen, J. S., & Brown, D. (2015). Why governments intervene: Exploring mixed motives for public policies on corporate social responsibility. *Public Policy and Administration, 30*(1), 51-72.

Kurtz, R. S. (2013). Oil spill causation and the deepwater horizon spill. *Review of Policy Research, 30*(4), 366-380.

Lester, R. K., & Hart, D. M. (2015). Closing the energy-demonstration gap. *Issues in Science and Technology, 31*(2), 48.

Lewis, D. (2001). The management of non-governmental development organizations: An introduction. In S. Van Puyvelde, R. Caers, C. Du Bois, & M. Jegers, M. (Eds.), *The governance of non-profit organizations: Integrating agency theory with stakeholder and stewardship theories.* London, UK: Routledge

Livesey, S. M. (2001). Eco-identity as discursive struggle: Royal Dutch/Shell, Brent Spar, and Nigeria. *Journal of Business Communication, 38*(1), 58-91.

Lynn, L. E., & Robichau, R. W. (2013). Governance and organisational effectiveness: Towards a theory of government performance. *Journal of Public Policy, 33*(2), 201.

Macondo. (2014). Investigation report overview: Explosion and fire at the Macondo well. *US Chemical Safety and Hazard Investigation Board, Report 2010-10-I-OS.*

Madubuko, C. (2014). Environment pollution: The rise of militarism and terrorism in the Niger Delta of Nigeria. *International Journal of Rural Law and Policy, 1,* 8.

Malterud, K. (2012). Systematic text condensation: A strategy for qualitative analysis. *Scandinavian Journal of Public Health, 40*(8), 795-805.

Mayowa, A. (2014). Pre-colonial Nigeria and the European's fallacy. *Review of History and Political Science, 2*(2), 17-27.

McGaghie, W. C., Bordage, G., & Shea, J. A. (2001). Problem statement, conceptual framework, and research question. *Journal of Academic Medicine, 76*(9).

Ministry of Power. (2016). *Official website of the Ministry of Power, Federal Republic of Nigeria.* Retrieved from http://www.power.gov.ng

National Petroleum Investment Management Services (NAPIMS). (2016). *Official website of NAPIMS.* Retrieved from http://www.napims.com

National Population Commission (NPC). (2018). *Official website of NPC.* Retrieved from http://www.population.com.ng

Niger Delta Development Commission (NDDC). (2016). *Official website of NDDC.* Retrieved from http://www.nddc.gov.ng

Nigerian Content Development and Monitoring Board (NCDMB). (2016). *Official website of NCDMB.* Retrieved from http://www.ncdmb.gov.ng

Nigerian Investment Promotion Commission (NIPC). (2016). *Official website of NIPC*. Retrieved from http://www.nipc.gov.ng

Nigeria Liquefied Natural Gas (NLNG). (2016). *Official website of NLNG*. Retrieved from http://www.nigerialng.com

Nigerian Maritime Administration and Safety Agency (NIMASA). (2016). *Official website of NIMASA*. Retrieved from http://www.nimasa.gov.ng

Nigerian National Petroleum Corporation (NNPC). (2016). *Official website for NNPC*. Retrieved from http://www.nnpcgroup.com

Nwaokoro, J. N. (2011). Nigeria's national content bill: The hype, the hope and the reality. *Journal of African Law, 55*(1), 128-155.

Nwapi, C. (2014). Enhancing the effectiveness of transparency in extractive resource governance: A Nigerian case study. *Law and Development Review, 7*(1), 23-47.

Nwoke, M. B. (2016). Relationship between natural economic resource and vocational choice among Nigeria youth: Psychological implications. *Asian Social Science, 12*(1), 84.

Nwokoma, N. I. (2015). Review, challenges, and future prospects of reforms in African economies: An appraisal of the Nigerian situation. *Journal of Economic Development, Management, IT, Finance, and Marketing, 7*(1), 1.

Obi, C. (2009). Nigeria's Niger Delta: Understanding the complex drivers of violent oil-related conflict. *Africa Development, 34*(2), 103-128.

Odoemene, A. (2012). The Nigerian armed forces and sexual violence in Ogoniland of the Niger Delta Nigeria, 1990-1999. *Armed Forces & Society, 38*(2), 225-251.

Oduntan, G. (2008). The emergent legal regime for exploration of hydrocarbons in the Gulf of Guinea: Imperative considerations for participating states and multinationals. *International and Comparative Law Quarterly, 57*(2), 253-302.

Ogbonna, G. N., & Ebimobowei, A. (2012). Petroleum income and Nigerian economy: Empirical evidence. *Arabian Journal of Business and Management Review, 1*(9), 33.

Ogunleye, E. K. (2008). Natural resource abundance in Nigeria: From dependence to development. *Resources Policy, 33*(3), 168-174.

Ojo, G. U. (2012). Community perception and oil companies' corporate social responsibility initiative in the Niger Delta. *Studies in Sociology of Science, 3*(4), 11-21.

Okeagu, J. E., Okeagu, J. C., Adegoke, A. O., & Onuoha, C. N. (2006). The environmental and social impact of petroleum and natural gas exploitation in Nigeria. *Journal of Third World Studies, 23*(1), 199-218.

Okonofua, B. A. (2016). The Niger Delta amnesty program: The challenges of transitioning from peace settlements to long-term peace. *Mediterranean Journal of Social Science, 6*(2), 1-16.

Okoye, A. (2012). Novel linkages for development: Corporate social responsibility, law and governance. Exploring the Nigerian petroleum industry bill. *Corporate Governance: The International Journal of Business in Society, 12*(4), 460-471.

Okpanachi, E. (2011). Confronting the governance challenges of developing Nigeria's extractive industry: Policy and performance in the oil and gas sector. *Review of Policy Research, 28*(1), 25-47.

Oluwatosin, A., Abimbola, O., & Olusegun, O. (2011). Oil price shocks and economic growth in Nigeria: Are thresholds important? *OPEC Energy Review, 35*(4), 208-333.

Omotola, J. S. (2009). "Liberation movements" and rising violence in the Niger Delta: The new contentious site of oil and environmental politics. *Studies in Conflict & Terrorism, 33*(1), 36-54.

Ordinioha, B., & Brisibe, S. (2013). The human health implications of crude oil spills in the Niger Delta, Nigeria: An interpretation of published studies. *Nigerian Medical Journal: Journal of the Nigeria Medical Association, 54*(1), 10.

Orji, U. J. (2014). Moving from gas flaring to gas conservation and utilization in Nigeria: A review of the legal and policy regime. *OPEC Energy Review, 38*(2), 149-183.

Orogun, P. S. (2010). Resource control, revenue allocation and petroleum politics in Nigeria: The Niger Delta question. *GeoJournal, 75*(5), 459-507.

O'Sullivan, E., Rassel, G. R., & Berner, M. (2008). *Research methods for public administrators.* Upper Saddle River, NJ: Pearson.

Ovadia, J. S. (2013). The Nigerian "one percent" and the management of national oil wealth through Nigerian content. *Science & Society, 77*(3), 315-341.

Pérouse de Montclos, M. (2014). The politics and crisis of the petroleum industry bill in Nigeria. *The Journal of Modern African Studies, 52*(3), 403-424.

Rodon, J., & Pastor, J. A. (2007). Applying grounded theory to study the implementation of an inter-organizational information system. *The Electronic Journal of Business Research Methods, 5*(2), 71-82.

Saidu, S., & Sadiq, H. A. (2014). Production sharing or joint venturing: What is the optimum petroleum contractual arrangement for the exploitation of Nigerian oil and gas? *Journal of Business and Management Sciences, 2*(2), 35-44.

Stringer, E. T. (2014). *Action research* (4th ed.). Thousand Oaks, CA: Sage Publications.

Swanson, D., Barg, S., Tyler, S., Venema, H., Tomar, S., Bhadwal, S., & Drexhage, J. (2010). Seven tools for creating adaptive policies. *Technological Forecasting & Social Change, 77*(6), 924-939.

Tracy, S. J. (2010). Qualitative quality: Eight "big-tent" criteria for excellent qualitative research. *Qualitative Inquiry, 16*(10), 837-851.

Ukpohor, E. T. (2013). Nigerian gas master plan: Strengthening the Nigeria gas infrastructure blueprint as a base for expanding regional gas market. *World Gas Conference Technical Paper, 1*(1), 1-18.

Ushie, V., Adeniyi, O., & Akongwale, S. (2013). Oil revenue, institutions and macroeconomic indicators in Nigeria. *OPEC Energy Review, 37*(1), 30-52.

Wagner, M. E., Alves, H., & Raposo, M. (2011). Stakeholder theory: Issues to resolve. *Management Decision, 49*(2), 226-252.

West African Gas Pipeline Company. (2016). *Official website for WAGP.* Retrieved from http://www.wagpco.com

Wheeler, D., Rechtman, R., Fabig, H., & Boele, R. (2001). Shell, Nigeria, and the Ogoni: A study in unsustainable development III. Analysis and implications of Royal Dutch/Shell group strategy. *Sustainable Development, 9*(4), 177-196.

Yardley, S. J., Watts, K. M., Pearson, J., & Richardson, J. C. (2014). Ethical issues in the reuse of qualitative data: Perspectives from literature, practice, and participants. *Qualitative Health Research, 24*(1), 102-113.

APPENDIX A

DEEPWATER GAS PRICING

QUESTIONNAIRE

Title: Deepwater gas pricing questionnaire (An action research questionnaire)

Research project title: Perspectives on monetizing the deepwater gas reserves in Niger Delta, Nigeria: An action research project

1. Establishing a stakeholder

    a. Do you have any affiliation to the Oil and Gas industry in Nigeria either by work, place of abode or peculiar interest? (Yes/No)

    b. Are you willing to share your perspectives about deepwater gas deposits in Niger Delta without any bias? (Yes/No)

c. Age (below 20, 21-30, 31-40,41-50, 51-60, above 60 years), gender (M/F), education (High School, graduate, post graduate)

2. Knowledge about oil and gas resource and income

a. About what percentage of Nigeria's foreign exchange earning comes from oil and gas (less than 25%, 26-50%, 51-75%, more than 75%)

b. About how much gas reserves does Nigeria have? (<100tcf, 100-200tcf, 200-300tcf, >300tcf)

c. About what percentage of the gas is located in deepwater Niger Delta (<1/3, 1/3-   2/3, >2/3)

3. Is there a government policy to exploit the deepwater gas for profit? (Yes/No/Don't know)

4. Government has the Nigerian Liquefied Natural Gas (NLNG) trains that have been exporting gas for profit. Do you think government can profitably monetize the deepwater gas for export at same pricing? (Yes/No/Don't know)

5. Government (in affiliation with her partners in the joint venture) has been utilizing gas from onshore for domestic uses. Do you think the deepwater gas can be utilized in a similar manner? (Yes/No/Don't know)

6. Gas from deepwater is typically more expensive to harness than onshore. Do you think the Nigerian government can give adequate incentives through monetization policy to help harness the gas profitably? (Yes/No/Don't know)

7. Do you see any connection between gas utilization from deepwater and industrialization? (Yes/No/Don't know)

8. Do you think the deepwater gas can help power the steel industry, the fertilizer and petrochemical plants? (Yes/No/Don't know)

9. Do you think Nigerian government will be willing to grant special incentives for deepwater gas monetization

to support growth and industrialization? (Yes/No/Don't know)

10. Do you think this kind of research with opinions from oil industry, government and community personalities could help government decide her monetization policies? (Yes/No/Don't know)

11. Do you think government can benefit from global experience of multinational oil companies in determining deepwater gas pricing? (Yes/No/Don't know)

12. Do you think there is real benefit to the communities from harnessing the deepwater gas deposits? (Yes/No/Don't know)

13. In your opinion, which of the three groups below will have the most benefit from harnessing the deepwater gas deposits?

    a. The multinational oil companies

    b. The government (federal/state/local)

c. The communities in the Delta

14. In your opinion, which of the three groups below will have the least benefit from harnessing the deepwater gas deposits?

    a. The multinational oil companies

    b. The government (federal/state/local)

    c. The communities in the Delta

15. The following questions will be open ended to solicit discussions on the opinions of the participants:

    a. What is your opinion about government policy on gas deposits in the Niger Delta?

    b. What do you think of the big oil companies' presence in deepwater Nigeria?

    c. What are your thoughts on treatment of host communities in the Niger Delta as it relates to oil and gas extraction?

    d. How well do you think the nation can utilize deepwater gas deposit for developmental purposes?

e. What other comment or remark do you have concerning the deepwater gas deposit in Niger Delta?

APPENDIX B

FOCUS GROUP

GUIDING QUESTIONS

There were five focus group questions to solicit open-ended conversations and discussions among the participants about monetizing deepwater gas in the Niger Delta.

1. What is your opinion about government policy on gas deposits in the Niger Delta? The question encouraged participants to freely express their opinions about gas policy.

2. What do you think of the big oil companies' presence in deepwater Nigeria? The concept of big oil companies relates to the multinationals that dominate the oil and gas industry in Nigeria. Their operating business philosophy sometimes

conflicts the realities of Nigerian business environment.

3. What are your thoughts on treatment of host communities in the Niger Delta as it relates to oil and gas extraction? The impact of oil and gas exploitation on the people and environment results in much activism, debate, and multifarious engagements in the Niger Delta.

4. How well do you think the nation can utilize deepwater gas deposit for developmental purposes? This question elicited responses concerning real value derivable from the deepwater gas.

5. What other comment or remark do you have concerning the deepwater gas deposit in Niger Delta? This open-ended question gathered any last thoughts or opinions from the participants that may not have been covered earlier.

www.ingramcontent.com/pod-product-compliance
Lightning Source LLC
Chambersburg PA
CBHW071256220526
45468CB00001B/149

* 9 7 8 1 7 9 0 3 1 3 4 4 0 *